Undergraduate Topics in Computer Science

Undergraduate Topics in Computer Science' (UTiCS) delivers high-quality instructional content for undergraduates studying in all areas of computing and information science. From core foundational and theoretical material to final-year topics and applications, UTiCS books take a fresh, concise, and modern approach and are ideal for self-study or for a one- or two-semester course. The texts are all authored by established experts in their fields, reviewed by an international advisory board, and contain numerous examples and problems. Many include fully worked solutions.

For other titles published in this series, go to
http://www.springer.com/7592

David Makinson

Sets, Logic and Maths
for Computing

 Springer

David Makinson, DPhil
London School of Economics, UK

Undergraduate Topics in Computer Science ISSN 1863-7310
ISBN: 978-1-84628-844-9   e-ISBN: 978-1-84628-845-6
DOI: 10.1007/978-1-84628-845-6

British Library Cataloguing in Publication Data
A catalogue record for this book is available from the British Library

Library of Congress Control Number: 2008927215

Springer Science+Business Media
springer.com

Preface

The first part of this preface is for the student; the second for the instructor. But whoever you are, welcome to both parts.

For the Student

You have finished secondary school, and are about to begin at a university or technical college. You want to study computing. The course includes some mathematics – and that was not necessarily your favourite subject. But there is no escape: some finite mathematics is a required part of the first year curriculum.

That is where this book comes in. Its purpose is to provide the basics – the essentials that you need to know to understand the mathematical language that is used in computer and information science.

It does not contain all the mathematics that you will need to look at through the several years of your undergraduate career. There are other very good, massive volumes that do that. At some stage you will probably find it useful to get one and keep it on your shelf for reference. But experience has convinced this author that no matter how good the compendia are, beginning students tend to feel intimidated, lost, and unclear about what parts to focus on. This short book, on the other hand, offers just the basics which you need to know from the beginning, and on which you can build further when needed.

It also recognizes that you may not have done much mathematics at school, may not have understood very well what was going on, and may even have grown to detest it. No matter: you can re-learn the essentials here, and perhaps even have fun doing so.

So what is the book about? It is about certain mathematical tools that we need to apply over and over again when working with computations. They include:

- *Collecting* things together. In the jargon of mathematics, this is *set theory*.

- *Comparing* things. This is the theory of *relations*.

- *Associating* one item with another. This is the theory of *functions*.

- *Recycling outputs as inputs*. We introduce the ideas of *induction* and *recursion*.

- *Counting*. The mathematician's term is *combinatorics*.

- *Weighing the odds*. This is done with the notion of *probability*.

- *Squirrel math*. Here we look at the use of *trees*.

- *Yea and Nay*. Just two truth-values underlie *propositional logic*.

- *Everything and nothing*. That is what *quantificational logic* is about.

Frankly, without an understanding of the basic concepts in these areas, you will not be able to acquire more than a vague idea of what is going on in computer science, nor be able to make computing decisions with discrimination or creatively. Conversely, as you begin to grasp them, you will find that their usefulness extends far beyond computing into many other areas of thought.

The good news is that there is not all that much to commit to memory. Your sister studying medicine, or brother going for law, will envy you terribly for this. In our subject, the essential thing in our kind of subject is to *understand*, and be able to *apply*.

But that is a much more subtle affair than you might imagine. Understanding and application are interdependent. Application without understanding is blind, and quickly leads to ghastly errors. On the other hand, comprehension remains poor without practice in application. In particular, you do not understand a definition until you have seen how it takes effect in specific situations: positive examples reveal its range, negative examples show its limits. It also takes time to recognize *when* you have really understood something, and when you have done no more than recite the words, or call upon it in hope of blessing.

For this reason, doing exercises is a indispensable part of the learning process. That is part of what is meant by the old proverb 'there is no royal road in mathematics'. It is also why we give so many problems and provide sample answers to some. Skip them at your peril: no matter how simple and straightforward a concept seems, you will not fully understand it unless you practice using it. So, even when an exercise is accompanied by a solution, you will benefit a great deal if you place a sheet over the answer and first try to work it out for yourself. That requires self-discipline and patience, but it brings real rewards.

In addition, the exercises have been chosen so that for many of them, the result is just what we need to make a step somewhere later in the book. They are thus integral to the development of the general theory.

By the same token, don't get into the habit of skipping the proofs when you read the text. In mathematics, you have never fully understood a fact unless you have also grasped *why* it is true, i.e. have assimilated at least one proof of it. The well-meaning idea that mathematics can be democratized by teaching the 'facts' and forgetting about the proofs has, in some countries, wrought disaster in secondary and university education in recent decades.

In practice, the mathematical tools that we bulleted above are rarely applied in isolation from each other. They gain their real power when used in combination, setting up a crossfire that can bring tough problems to the ground. The concept of a set, once explained in the first chapter, is used absolutely everywhere in what follows. Relations reappear in graphs and trees. The familiar arithmetical operations of addition and multiplication, heavily employed in combinatorics and probability, are of course particular examples of functions. And so on.

For the Instructor

Any book of this kind needs to find a delicate balance between the competing demands of intrinsic mathematical order and those of intuition. Mathematically, the most elegant and coherent way to proceed is to begin with the most general concepts, and gradually unfold them so that the more specific and familiar ones appear as special cases. Pedagogically, this sometimes works, but it can also be disastrous. There are situations where the reverse is often required: begin with some of the more familiar special cases, and then show how they may naturally be broadened into cover much wider terrain.

There is no perfect solution to this problem; we have tried to find a least imperfect one. Insofar as we begin the book with sets, relations and functions in that order, we are following the first path. But in some chapters we have followed the second one. For example, when explaining induction and recursion we begin with the most familiar special case, simple induction/recursion over the positive integers; passing to their cumulative forms over the same domain; broadening to their qualitatively formulated structural versions; and finally presenting the most general forms on arbitrary well-founded sets. Again, in the chapter on trees, we have taken the rather unusual step of beginning with rooted trees, where intuition is strongest, then abstracting to unrooted trees.

In the chapters on counting and probability we have had to strike another balance – between traditional terminology and notation, which antedates the modern era, and its translation into the language of sets, relations and functions. Most textbook presentations do everything the traditional way, which has its drawbacks. It leaves the student in the dark about the relation of this material to what was taught in earlier chapters on sets, relations and functions. And, frankly, it is not always very rigorous or transparent. Our policy is to familiarize the reader with *both* kinds of presentation – using the language of sets and functions for a clear understanding of the material itself, and the traditional languages of combinatorics and probability to permit communication in the local dialect.

The place of logic in the story is delicate. We have left its systematic exposition to the end, a decision that may seem rather strange. For surely one uses logic whenever reasoning mathematically, even about such elementary things as sets, relations and functions, covered in the first three chapters. Don't we need a chapter on logic at the very beginning? The author's experience in the classroom tells him that in practice that does not work well. Despite its simplicity – indeed because of it – logic can appear intangible for beginning students. It acquires intuitive meaning only as its applications are revealed. Moreover, it turns out that a really clear explanation of the basic concepts of logic requires some familiarity with the mathematical notions of sets, relations, functions and trees.

For these reasons, the book takes a different tack. In its early chapters, notions of logic are identified briefly as they arise in the discussion of more 'concrete' material. This is done in 'logic boxes'. Each box introduces just enough to get on with the task in hand. Much later, in the last two chapters, all this is brought together and extended in a systematic presentation. By then, the student will have little trouble appreciating what the subject is all about, and how natural it all is.

From time to time there are boxes of a different nature – 'Alice boxes'. This little trouble-maker comes from the pages of Lewis Carroll, to ask embarrassing questions in all innocence. Often they are questions that bother students, but which they have trouble articulating clearly or are too shy to pose. In particular, it is a fact that the house of mathematics and logic can be built in many different ways, and sometimes the constructions of one text appear to be in conflict with those of another. Perhaps the most troublesome example of this comes up in quantificational logic, with different ways of reading the quantifiers and even different ways of using the terms 'true' and 'false'. But there are also plenty of others, in all the chapters. It is hoped that the Mad Hatter's responses are of assistance.

Overall, our choice of topics is fairly standard, as the chapter titles indicate. If strapped for class time, an instructor could omit some of the later sections of Chapters 5–9, perhaps even from the end of Chapter 4. But it is urged that

Chapters 1–3 be kept intact, as everything in them is subsequently needed. Some instructors may wish to add or deepen topics; as this text makes no pretence to cover all possible areas of focus, such extensions are left to their discretion.

We have not included a chapter on the theory of graphs. This was a difficult call to make, and the reasons were as follows. Although trees are a particular kind of graph, there is no difficulty in covering everything we want to say about trees, without entering into the more general theory. Moreover, an adequate treatment of graphs, even if squeezed into one chapter of about the same length as the others, takes a good two weeks of additional class time to cover properly. The general theory of graphs is a rather messy area, with options about how wide to cast the net (graphs with or without loops, multiple edges etc as well as the basic distinction between directed and undirected graphs), and a rather high definition/theorem ratio. The author's experience is that students gain little from a high-speed run through these distinctions and definitions, memorized for the examinations and then promptly forgotten.

Finally, a decision had to be made whether to include specific algorithms and, if so, in what form: ordinary English, pseudo-code outline, or a real-life programming language in full detail? Our decision has been to leave options open as much as possible. In principle, most first year students of computing will be taking, in parallel, courses on principles of programming and some specific programming language. But the programming languages chosen will differ from one institution to another. The policy in this text is to sketch the essential idea of basic algorithms in plain but carefully formulated English. In some cases (particularly the chapter on trees), we give optional exercises in expressing them in pseudo-code. Instructors wishing to make more systematic use of pseudo-code, or to link material with specific programming languages, should feel free to do so.

Acknowledgements

The author would like to thank Anatoli Degtyarev, Franz Dietrich, Valentin Goranko and George Kourousias for helpful discussions. George also helped prepare the diagrams. Thanks as well to the London School of Economics (LSE), and particularly department heads Colin Howson and Richard Bradley, for the wonderful working environment that they provided while this book was being written. Thanks as well to the London School of Economics(LSE), and particularly department heads Colin Howson and Richard Bradley, for the wonderful working environment that they provided while this book was being written.

Contents

1

Collecting Things Together: Sets

Chapter Outline

In this chapter we introduce the student to the world of sets. Actually, only a little bit of it, the part that is needed to get going.

After giving a rough intuitive idea of what sets are, we present the basic relations between them: *inclusion*, *identity*, *proper inclusion*, and *exclusion*. We describe two common ways of identifying sets, and pause to look more closely at the *empty set*. We then define some basic operations for forming new sets out of old ones: *intersection*, *union*, *difference* and *complement*. These are often called Boolean operations, after George Boole who first studied them systematically in the middle of the nineteenth century.

Up to this point, the material is all 'flat' set theory, in the sense that it does not look at what happens when we build sets of sets. However we need to go a little beyond the flat world. In particular, we generalize the notions of intersection and union to cover arbitrary *collections of sets*, and introduce the very important concept of the *power set* of a set, i.e. the set of all its subsets.

1.1 The Intuitive Concept of a Set

Every day you need to consider things more than one at a time. As well as thinking about a particular individual, such as the young man or woman sitting on your left in the classroom, you may focus on some collection of people – say, all those

D. Makinson, *Sets, Logic and Maths for Computing*,
DOI: 10.1007/978-1-84628-845-6_1, © Springer-Verlag London Limited 2008

students who come from the same school as you do, or all those with red hair. A *set* is just such a collection, and the individuals that make it up are called its *elements*. For example, each student with green eyes is an element of the set of all students with green eyes.

What could be simpler? But be careful! There might be many students with green eyes, or none, or maybe just one, but there is always exactly one *set* of them. It is a single item, even when it has many elements. Moreover, whereas the students themselves are flesh-and-blood persons, the set is an abstract object, thus different from those elements. Even when the set has just one element, it is not the same thing as that unique element. For example, even if Achilles is the only person in the class with green eyes, the corresponding set is distinct from Achilles; it is an abstract item and not a person. To anticipate later terminology, the point is often marked by calling it the *singleton* for that person, and writing it as {Achilles}.

The elements of a set need not be people. They need not even be physical objects; they may in turn be abstract items. For example, they can be numbers, geometric figures, items of computer code, colours, concepts, or whatever you like. . .and even other sets.

We need a notation to represent the idea of elementhood. We write $x \in A$ for x is an element of A, and $x \notin A$ for x is *not* an element of A. Here, A is a set; x may or may not be a set; in the simple examples it will not be one. The sign $\in$ is derived from one of the forms of the Greek letter epsilon.

1.2 Basic Relations Between Sets

1.2.1 Inclusion

Sets can stand in various relations to each other. One basic relation is that of *inclusion*. When A, B are sets, we say that A is *included in B* (or: A is a *subset of B*) and write $A \subseteq B$ iff every element of A is an element of B. In other words, iff for all x, if $x \in A$ then $x \in B$. Put in another way that is sometimes useful, iff there is no element of A that is not an element of B. Looking at the same relation from the other side, when this holds, we also say that B *includes A* (B is a *superset* of A) and write $B \supseteq A$.

Alice Box: iff

Alice: Hold on, what's this 'iff'? It's not in my dictionary.

Hatter: Too bad for your dictionary. The expression was introduced around the middle of the last century by the mathematician Paul Halmos, as a handy

(Continued)

Alice Box: (Continued)

shorthand for 'if and only if', and soon became standard among mathematicians.

Alice: OK, but aren't we doing some *logic* here? I see words like 'if', 'only if', 'every', 'not', and perhaps more. Shouldn't we begin by explaining what they mean?

Hatter: We could, but life will be easier if for the moment we simply use these particles as you would in everyday life. We will get back to their exact logical analysis later.

EXERCISE 1.2.1 (WITH SOLUTION)

Which of the following sets are included in which? Use the notation above, and express yourself as succinctly and clearly as you can. Recall that a prime number is a positive integer greater than 1 that is not divisible by any positive integer other than itself and 1.

A : The set of all positive integers less than 10

B : The set of all prime numbers less than 11

C : The set of all odd numbers greater than 1 and less than 6

D : The set whose only elements are 1 and 2

E : The set whose only element is 1

F : The set of all prime numbers less than 8

Solution: Each of these sets is included in itself, and each of them is included in A. In addition, we have $C \subseteq B$, $E \subseteq D$, $F \subseteq B$, $B \subseteq F$.

Comments: Note that none of the other converses hold. For example, we do not have $B \subseteq C$, since $7 \in B$ but $7 \notin C$. Note also that we do not have $E \subseteq B$ since 1 is not a prime number.

Warning: Avoid saying that A is 'contained' in B, as this is rather ambiguous. It can mean that $A \subseteq B$, but it can also mean that $A \in B$. *These are not the same, and should never be confused.* For example, the integer 2 is an element of the set $\mathbf{N}^+$ of all positive integers, but it is not a subset of $\mathbf{N}^+$. Conversely, the set E of all even integers is a subset of $\mathbf{N}^+$, i.e. each of its elements 2, 4,... is an element of $\mathbf{N}^+$; but E is not an element of $\mathbf{N}^+$.

1.2.2 Identity

The notion of inclusion leads us to the concept of *identity* (alias equality) between sets. Clearly, if both $A \subseteq B$ and $B \subseteq A$ then A and B have exactly the same elements – every element of either is an element of the other; in other words, there is no element of one that is not an element of the other. A basic principle of set theory, called the axiom (or postulate) of *extensionality*, says something more: *When both $A \subseteq B$ and $B \subseteq A$ then the sets A, B are in fact identical.* They are one and the same set, and we write $A = B$.

EXERCISE 1.2.2 (WITH SOLUTION)

Which sets in the top row are identical with their counterparts in the bottom row? Read the curly braces as framing the elements of the set so that, for example, $\{1, 2, 3\}$ is the set whose elements are just 1, 2, 3. Be careful with the answers.

$\{1, 2, 3\}$	$\{9, 5\}$	$\{0, 2, 8\}$	$\{7\}$	$\{8\}$	$\{$London, Leeds$\}$
$\{3, 2, 1\}$	$\{9, 5, 9\}$	$\{\lfloor\sqrt{4}\rfloor, 0/5, 2^3\}$	7	$\{\{8\}\}$	$\{$Londres, 'Leeds'$\}$

Solution:

yes	yes	yes	no	no	no

Comments:

Column 1: The order of enumeration makes no difference – the sets still have the same elements.

Column 2: Repeating an element in the enumeration is inelegant, but it makes no difference – the sets still have the same elements.

Column 3: The elements have been named differently, as well as being written in a different order but they are the same.

Column 4: 7 is a number, not a set, while $\{7\}$ is a set with the number 7 as its only element, i.e. its singleton.

Column 5: This time, top and bottom are both sets, and they both have just one element, but these elements are not the same. The unique element of the top set is the number 8, while the unique element of the bottom set is the set $\{8\}$. The bottom set is the singleton of the top one.

Column 6: These are both two-element sets. The first-mentioned elements are the same: London is the same city as Londres, although they are named in different languages. But the second-mentioned elements are not the same: Leeds is a city whereas 'Leeds' is the name of that city.

The distinctions in the last three columns may seem rather pedantic, but they turn out to be very important to avoid confusions when we are dealing with sets of sets or with sets of symbols, as is often the case in computer science.

EXERCISE 1.2.3 (WITH SOLUTION)

In Exercise 1.2.1, which of the sets are identical to which?

Solution: $B = F$ (so that also $F = B$). And of course $A = A$, $B = B$ etc – each of the listed sets is identical to itself.

Comment: The fact that we defined B and F in different ways makes no difference: the two sets have exactly the same elements and so by the axiom of extensionality are identical.

1.2.3 Proper Inclusion

When $A \subseteq B$ but $A \neq B$ then we say that A is *properly* included in B, and write $A \subset B$. Sometimes $\subset$ is written with a small $\neq$ underneath. That should not cause any confusion, but another notational dialect is more dangerous: a few older texts use $A \subset B$ for plain inclusion. Be wary when you read.

EXERCISE 1.2.4 (WITH SOLUTION)

In Exercise 1.2.1, which of the sets are properly included in which? In each case give a 'witness' to the proper nature of the inclusion, i.e. identify an element of the right one that is not an element of the left one.

Solution: $C \subset B$, witnesses 2,7; $E \subset D$, sole witness 2.

Comment: $B \not\subset F$, since B and F have exactly the same elements.

1.2.4 Euler Diagrams

If we think of a set A as represented by all the points in a circle (or other closed plane figure) then we can represent the notion of one set A being a proper subset of another B by putting a circle labelled A inside a circle labelled B. We can diagram equality, of course, by drawing just one circle and labelling it both A and B. Thus we have the following *Euler diagrams* (so named after the eighteenth century mathematician Euler who used them when teaching a princess by correspondence):

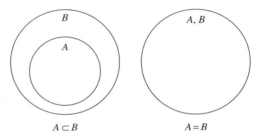

Figure 1.1   *Euler diagrams for proper inclusion and identity.*

How can we diagram inclusion in general? Here we must be careful. *There is no single Euler diagram that does the job.* When $A \subseteq B$ then we may have either of the above two configurations: if $A \subset B$ then the left diagram is appropriate, if on the other hand $A = B$ then the right diagram is the correct one.

Diagrams are a very valuable aid to intuition, and it would be pedantic and unproductive to try to do without them. But we must also be clearly aware of their limitations. If you want to visualize $A \subseteq B$ and you don't know whether the inclusion is proper, you will need to consider two Euler diagrams and see what happens in each.

1.2.5 Venn Diagrams

Alternatively, you can use another kind of diagram, which can represent plain inclusion without ambiguity. It is called a *Venn diagram* (after the nineteenth century logician John Venn). It consists of drawing two circles, one for A and one for B, always intersecting no matter what the relationship between A and B, and then putting a mark (e.g. $\varnothing$) in an area to indicate that it has no elements, and another kind of mark (e.g. a cross) to indicate that it does have at least one element. With these conventions, the left diagram below represents $A \subseteq B$ while the right diagram represents $A \subset B$.

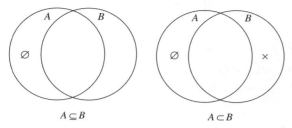

Figure 1.2 *Venn diagrams for inclusion and proper inclusion.*

Note that the disposition of the circles is always the same: what changes are the areas noted as empty or as non-empty. There is considerable variability in the signs used for this, dots, ticks, etc.

A great thing about Venn diagrams is that they can represent common relationships like $A \subseteq B$ by a single diagram, rather than by an alternation of different diagrams. Another advantage is that with some additions they can be used to represent basic operations on sets as well as relations between them. The bad news is that when you have more than two sets to consider, Venn diagrams quickly become very complicated and lose their intuitive clarity – which was, after all, their principal *raison d'être*.

Warning: Do not confuse Euler diagrams with Venn diagrams. They are constructed differently and read differently. Unfortunately, textbooks themselves are sometimes sloppy here, using the terms interchangeably or even in reverse.

1.2.6 Ways of Defining a Set

The work we have done so far already illustrates two important ways of defining or identifying a set. One way is to *enumerate all its elements individually* between curly brackets, as we did in Exercise 1.2.2. Evidently, such an enumeration can be completed only when there are finitely many elements, and in practice only when the set is fairly small. The order of enumeration makes no difference to what items are elements of the set, e.g. $\{1,2,3\} = \{3,1,2\}$; but we usually write elements in some conventional order such as increasing size to facilitate reading.

Another way of identifying a set is by *providing a common property*: the elements of the set are understood to be all (and only) the items that have that property. That is what we did for several of the sets in Exercise 1.2.1. There is a notation for this. For example, we write the first set as follows: $A = \{x \in \mathbf{N}^{+}:$ $x < 10\}$. Here $\mathbf{N}^{+}$ stands for the set of all integers greater than zero (the positive integers). Some authors use a vertical bar in place of a colon.

EXERCISE 1.2.5 (WITH SOLUTION)

(a) Identify the sets A, B, C, F of Exercise 1.2.1 by enumeration.

(b) Identify the sets D, E of the same exercise by properties, using the notation introduced.

Solution:

(a) $A = \{1,2,3,4,5,6,7,8,9\}$; $B = \{2,3,5,7\}$; $C = \{3,5\}$, $F = \{2,3,5,7\}$.

(b) There are many ways of doing this, here are some. $D = \{x \in \mathbf{N}^{+}: x$ divides all even integers$\}$; $E = \{x \in \mathbf{N}^{+}: x$ is less than or equal to every positive integer$\}$.

When a set is infinite, we often use an incomplete 'suspension points' notation. Thus, we might write the set of all even positive integers and the set of all primes respectively as follows: $\{2, 4, 6,...\}$, $\{2, 3, 5, 7, 11, ...\}$. But it should be emphasized that this is an informal way of writing, used when it is well understood between writer and reader what particular continuation is intended. Clearly, there are many ways of continuing each of these partial enumerations. We normally understand that the most familiar or simplest is the one that is meant.

These two methods of identifying a set – by enumeration and by a common property – are not the only ones. In a later chapter we will be looking at another very important one, known as recursive definition.

EXERCISE 1.2.6 (WITH PARTIAL SOLUTION)

True or false? In each case use your intuition to make a guess, and establish it by either proving the point from the definitions (if you guessed positively) or giving a simple counterexample (if you guessed negatively). Make sure that you don't confuse $\subseteq$ with $\subset$.

(a) Whenever $A \subseteq B$ and $B \subseteq C$ then $A \subseteq C$

(b) Whenever $A \subseteq B$ and $C \subseteq B$ then $A \subseteq C$

(c) Whenever $A_1 \subseteq A_2 \subseteq ... \subseteq A_n$ and also $A_n \subseteq A_1$ then $A_i = A_j$ for all $i,j \le n$

(d) $A \subset B$ iff $A \subseteq B$ and $B \nsubseteq A$

(e) $A \subseteq B$ iff $A \subset B$ or $A = B$

(f) $A = B$ iff neither $A \subset B$ nor $B \subset A$

(g) Whenever $A \subseteq B$ and $B \subset C$ then $A \subset C$

(h) Whenever $A \subset B$ and $B \subseteq C$ then $A \subset C$

Solutions to (a), (b), (f), (h):

(a) True. Take any sets A, B, C. Suppose $A \subseteq B$ and $B \subseteq C$; it suffices to show $A \subseteq C$. Take any x, and suppose $x \in A$; by the definition of inclusion, it is enough to show $x \in C$. But since $x \in A$ and $A \subseteq B$ we have by the definition of inclusion that $x \in B$. So since also $B \subseteq C$ we have again by the definition of inclusion that $x \in C$, as desired.

(b) False. Counterexample: $A = \{1\}$, $B = \{1,2\}$, $C = \{2\}$.

(f) False. The left to right implication is correct, but the right to left one is false, so that the entire co-implication (the 'iff') is also false. Counter-example: $A = \{1\}$, $B = \{2\}$.

(h) True. Take any sets A, B, C. Suppose $A \subset B$ and $B \subseteq C$. From the former by the definition of proper inclusion we have $A \subseteq B$. So by exercise (a) we have $A \subseteq C$. It remains to show that $A \neq C$. Since $A \subset B$ we have by exercise (d) that $B \not\subseteq A$, so by the definition of inclusion there is an x with $x \in B$ but $x \notin A$. Thus since $B \subseteq C$ we have $x \in C$ while $x \notin A$, so that $C \not\subseteq A$ and thus $A \neq C$ as desired.

Comment: The only false ones are (b) and (f). All the others are true.

We have given the proofs of the positive solutions in quite full detail, perhaps even to the point of irritation. The reason is that they illustrate some general features of *proof construction*, which we now articulate.

Logic Box: Proving general and conditional statements

Proving general statements. If you want to prove a statement about all things of a certain kind, a straightforward line of attack is to consider an arbitrary one of those things, and show that the statement holds of it. In the example, we wanted to show that *whenever $A \subseteq B$ and $B \subseteq C$ then $A \subseteq C$*. We did this by *choosing arbitrary A,B,C, and working with them.* This procedure is so obvious that can pass unnoticed; indeed, we often won't bother to mention it explicitly. But the logical principles underlying the procedure are important and quite subtle, as we will see in a later chapter when we discuss the ideas of universal instantiation and generalization.

Proving conditional statements. If you want to prove a statement of the form 'if this then that', a straightforward line of attack is to *suppose that the*

(Continued)

Logic Box: (Continued)

first is true, and on that basis *show that the second is true.* We did this in the proof of (a). Actually, we did it twice : first we supposed that $A \subseteq B$ and $B \subseteq C$ were true, and set our goal as showing $A \subseteq C$. Later, we supposed that $x \in A$, and aimed to get $x \in C$.

Our examples also illustrate some *heuristics* (rough guides) for finding proofs. They are hardly more than commonsense, but when overlooked can be a source of failure and confusion.

Proof heuristics: Destination, starting point, toolkit

Always be clear what you are trying to show. If you don't know what you are trying to prove it is unlikely that you will prove it, and if by chance you do, the proof will probably be a real mess. Following this rule is not as easy as may appear, for *as a proof develops, the goal changes*! For example, in Exercise 1.2.6 (a), we began by trying to show (a) itself. After choosing A, B, C arbitrarily, we sought to prove the conditional statement 'If $A \subseteq B$ and $B \subseteq C$ then $A \subseteq C$'. Then, after supposing that $A \subseteq B$ and $B \subseteq C$, we switched our goal to $A \subseteq C$. We then chose an arbitrary x, supposed $x \in A$, and aimed for $x \in C$. In half a dozen lines, four different goals! At each stage we have to be aware of which one we are driving at. When we start using more sophisticated tools for building proofs, such as argument via contraposition and *reductio ad absurdum* (to be explained later), the goal-shifts become even more striking.

As far as possible, be aware of what you are allowed to use, and don't hesitate to use it. What are you allowed to use? In the first place, you may use the *definitions* of terms in the problem (in our example, the notions of subset and proper subset). Too many students come to mathematics with the idea that a definition is just something for decoration, something that you can hang on the wall like a picture or diploma. A definition is for use. In a very simple proof, half the steps can consist of 'unpacking' the definitions and then, after reasoning, packing them together again. In the second place, you may use whatever basic *axioms* (alias postulates) that you have been supplied with. In the exercise, that was just the principle of extensionality. In the third place, you can use anything that you have *already proven.* In the exercise, we did this while proving (h).

Be flexible and ready to go into reverse. If you can't prove that a statement is true, try looking for a counterexample in order to show that it is false. If you can't find a counterexample, try to prove that it is true. With some experience,

(*Continued*)

> *Proof heuristics:* (Continued)
>
> you can often use the failure of your attempted proofs as a guide to finding a suitable counterexample, and the failures of your trial counterexamples to give a clue for constructing a proof. This is all part of the *art of proof and refutation.*

1.3 The Empty Set

1.3.1 Emptiness

What do the following two sets have in common?

$$A = \{x \in \mathbf{N}^+ : x \text{ is both even and odd}\}$$
$$B = \{x \in \mathbf{N}^+ : x \text{ is prime and } 24 \leq x \leq 28\}$$

Answer: neither of them has any elements. From this it follows that they have exactly the same elements – neither has any element that is not in the other. So by the principle of extensionality, they are identical, i.e. they are the same set. Thus $A = B$, even though they are described differently. This leads to the following the definition. *The empty set, written $\varnothing$, is defined to be the (unique) set that has no elements at all.*

This is a very important set, just as zero is a very important number. The following exercise gives one of its basic properties.

EXERCISE 1.3.1 (WITH SOLUTION)

Show that $\varnothing \subseteq A$ for every set A.

Solution: We need to show that for all x, if $x \in \varnothing$ then $x \in A$. In other words: there is no x with $x \in \varnothing$ but $x \notin A$. But by the definition of $\varnothing$, there is no x with $x \in \varnothing$, so we are done.

Alice Box: if. . .then. . .

Alice: That's a short proof, but a strange one. You say 'in other words', but are the two formulations really equivalent?

Hatter: Indeed they are. This is because of the way in which we understand 'if. . .then. . .' statements in mathematics. We could explain that in detail now, but it is probably better to come back to it a bit later.

Alice: It's a promise?

Hatter: It's a promise!

1.3.2 Disjoint Sets

With the empty set in hand, we can define a final relation between sets. We say that sets A, B are *disjoint* (alias mutually exclusive) iff they have no elements in common. That is, iff there is no x such that both $x \in A$ and $x \in B$. When they are not disjoint, i.e. have at least one element in common, we can say that they *overlap*.

More generally, when $A_1, \ldots, A_n$ are sets, we say that they are *pairwise disjoint* iff for any $i,j \leq n$, if $i \neq j$ then A_i has no elements in common with A_j.

EXERCISE 1.3.2

  (a) Of the sets in Exercise 1.2.1, which are disjoint from which?

  (b) Draw a Euler diagram and also a Venn diagram to express the situation that A and B are disjoint. Draw a Venn diagram to expres the situation that they overlap. Why is there no single Euler diagram for the latter?

  (c) Construct three sets X, Y, Z such that X is disjoint from Y and Y is disjoint from Z, but X is not disjoint from Z.

  (d) Show that the empty set is disjoint from every set, including itself.

1.4 Boolean Operations on Sets

We now define some operations on sets, that is, ways of constructing new sets out of old. There are three basic ones: intersection, meet, and relative complement, and several others that can be defined in terms of them.

1.4.1 Intersection

If A and B are sets, we define their *intersection* $A \cap B$, also known as their *meet*, by the following rule. For all x:

$$x \in A \cap B \text{ iff } x \in A \text{ and } x \in B$$

EXERCISE 1.4.1 (WITH PARTIAL SOLUTION)

  Show the following :

  (a) $A \cap B \subseteq A$ and $A \cap B \subseteq B$

  (b) Whenever $X \subseteq A$ and $X \subseteq B$ then $X \subseteq A \cap B$

(c) $A \cap B = B \cap A$ (commutation principle)

(d) $A \cap (B \cap C) = (A \cap B) \cap C$ (association)

(e) $A \cap A = A$ (idempotence)

(f) $A \cap \varnothing = \varnothing$ (bottom)

(g) Reformulate the definition of disjoint sets using intersection.

Solutions to (b), (f), (g):

For (b): Suppose $X \subseteq A$ and $X \subseteq B$; we want to show $X \subseteq A \cap B$. Take any x and suppose $x \in X$; we need to show that $x \in A \cap B$. But since $x \in X$ and $X \subseteq A$ we have by the definition of inclusion that $x \in A$; and similarly since $x \in X$ and $X \subseteq B$ we have $x \in B$. So by the definition of intersection, $x \in A \cap B$ as desired.

For (f): We already have $A \cap \varnothing \subseteq \varnothing$ by (a) above. And we also have $\varnothing \subseteq A \cap \varnothing$ by Exercise 1.3.1, so we are done.

For (g): Sets A, B are disjoint iff $A \cap B = \varnothing$.

Logic Box: Conjunction

Intersection is defined using the word 'and'. But what does this mean? In mathematics it is very simple – much simpler than in ordinary life. Consider any two statements (alias propositions) α, β. Each can be true, or false, but not both. When is the statement 'α and β', called the *conjunction* of the two parts, true? The answer is intuitively clear: when each of α, β considered separately is true, the conjunction is true, but in all other cases the conjunction is false. What are the other cases? There are three of them: α true with β false, α false with β true, α false with β false.

What we have just said can be put in the form of a table, called the *truth-table for conjunction*.

α	β	$\alpha \wedge \beta$
1	1	1
1	0	0
0	1	0
0	0	0

To read this table, each row represents a possible combination of truth-values of the parts α, β. For brevity we write 1 for 'true' and 0 for 'false'. The rightmost entry in the row gives us the resulting truth-value of the conjunction 'α and β', which we write as $\alpha \wedge \beta$. Clearly, the truth-value of the conjunction is fully

(Continued)

Alice Box: (Continued)

determined by each combination of truth-values of the parts. For this reason, conjunction is called a *truth-functional logical connective*.

In a chapter on logic, we will look at the properties and behaviour of conjunction. As you may already have guessed, the behaviour of intersection as an operation on sets reflects that of conjunction as a connective between propositions. This is because the latter is used in the definition of the former. For example, the commutativity of intersection is a reflection of the fact that 'α and β' has exactly the same truth-conditions as 'β and α'.

For reflection: How do you square this with the difference in meaning between 'They got married and had a baby' and 'They had a baby and got married'?

1.4.2 Union

Alongside intersection we have another operation called *union*. The two operations are known as *duals* of each other, in the sense that each is like the other 'upside down'. For any sets A and B, we define their union $A \cup B$ by the following rule. For all x:

$$x \in A \cup B \text{ iff } x \in A \text{ or } x \in B,$$

where this is understood in the sense:

$$x \in A \cup B \text{ iff } x \in A \text{ or } x \in B \text{ (or both)},$$

in other words:

$$x \in A \cup B \text{ iff } x \text{ is an element of } at\ least\ one \text{ of } A,\ B.$$

The contrast with intersection may be illustrated by Venn diagram. The two circles represent the sets A, B. The left shaded area represents $A \cup B$, while the right shaded area represents $A \cap B$.

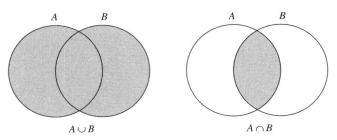

Figure 1.3 *Venn diagrams for union and intersection.*

The properties of union are just like those of intersection but 'upside down'. Evidently, this is a rather vague way of speaking; it can be made precise, but it is better to leave the idea on an intuitive level for the moment.

EXERCISE 1.4.2

Show the following:

(a) $A \subseteq A \cup B$ and $B \subseteq A \cup B$

(b) Whenever $A \subseteq X$ and $B \subseteq X$ then $A \cup B \subseteq X$

(c) $A \cup B = B \cup A$ (commutation principle)

(d) $A \cup (B \cup C) = (A \cup B) \cup C$ (association)

(e) $A \cup A = A$ (idempotence)

(f) $A \cup \varnothing = A$ (bottom)

Logic Box: Disjunction

When 'or' is understood in the sense that we have described, it is known as (inclusive) *disjunction* and statements 'α or β' are written as $\alpha \vee \beta$. Whereas there is just one way (out of four) of making a conjunction true, there is just way of making a disjunction false. The truth-table is as follows:

α	β	$\alpha \vee \beta$
1	1	1
1	0	1
0	1	1
0	0	0

Clearly, the truth-value of the disjunction is fully determined by each combination of truth-values of the parts. In other words, it is also a truth-functional logical connective.

The behaviour of union between sets reflects that of disjunction as a connective between propositions.

For reflection: In ordinary discourse we often use 'α or β' to mean 'either α, or β, but not both' i.e. 'exactly one of α, β is true'. This is called *exclusive disjunction*. What would its truth-table look like?

The last two exercises set out some of the basic properties of intersection and of union, taken separately. But how do they relate to each other? The following exercise covers the most important interactions.

EXERCISE 1.4.3 (WITH PARTIAL SOLUTION)

Show the following:

(a) $A \cap B \subseteq A \cup B$

(b) $A \cap (A \cup B) = A = A \cup (A \cap B)$ (absorption)

(c) $A \cap (B \cup C) = (A \cap B) \cup (A \cap C)$ (distribution of intersection over union)

(d) $A \cup (B \cap C) = (A \cup B) \cap (A \cup C)$ (distribution of union over intersection)

Solution to (c): Writing LHS, RHS for the left and right hand sides respectively, we need to show that LHS $\subseteq$ RHS and conversely RHS $\subseteq$ LHS.

For LHS $\subseteq$ RHS, suppose that $x \in$ LHS. Then $x \in A$ and $x \in B \cup C$. From the latter we have that either $x \in B$ or $x \in C$. Consider the two cases separately. Suppose first that $x \in B$. Since also $x \in A$ we have $x \in A \cap B$ and so by Exercise 1.4.2 (a), $x \in (A \cap B) \cup (A \cap C) =$ RHS as desired. Suppose second that $x \in C$. Since also $x \in A$ we have $x \in A \cap C$ and so again $x \in (A \cap B) \cup (A \cap C) =$ RHS as desired.

For RHS $\subseteq$ LHS, suppose that $x \in$ RHS. Then $x \in A \cap B$ or $x \in A \cap C$. Consider the two cases separately. Suppose first that $x \in A \cap B$. Then $x \in A$ and $x \in B$; from the latter $x \in B \cup C$, and so with the former, $x \in A \cap (B \cup C) =$ LHS as desired. Suppose second that that $x \in A \cap C$. The argment is similar: $x \in A$ and $x \in C$; from the latter $x \in B \cup C$, and so with the former, $x \in A \cap (B \cup C) =$ LHS as desired.

Logic Box: Proof by cases

In the exercise above we used a technique known as proof by cases, or disjunctive proof. Suppose we know that either α is true or β is true, but we don't know which. It can be difficult to proceed with this rather weak information. So we break the argument into two parts.

First we suppose that α is true (the first case) and with this stronger assumption we head for whatever it was that we were trying to establish.

Then we suppose instead that β is true (the second case) and argue using this assumption to the same conclusion.

(Continued)

Alice Box: (Continued)

In each case the goal remains unchanged, but the two cases must be treated quite separately: we cannot use the supposition for the first case in the argument for the second case, and vice versa. If we succeed in reaching the desired conclusion in each case separately, then we know that it must hold *irrespective of which case is true.* The arguments carried out in the two separate cases may sometimes resemble each other closely (as in our exercise). But in more challenging problems they may be very different.

Alice Box: Overlapping cases

Alice: What if *both* cases are true? For example, in the solution to the preceding exercise, in say the part for LHS $\subseteq$ RHS: what if both $x \in B$ and $x \in C$?

Hatter: No problem! This just means that we have covered that situation twice. For proof by cases to work, it is not required that the two cases be exclusive. In some examples (as in our exercise) it is easier to work with overlapping cases; sometimes it is more elegant and economical to work with cases that exclude each other.

1.4.3 Difference and Complement

There is one more Boolean operation on sets that we wish to consider: *difference.* Let A, B be any sets. We define the *difference of B in A*, written $A \backslash B$ (also as $A-B$) to be the set of all elements of A that are *not* elements of B. That is, $A \backslash B = \{x : x \in A \text{ but } x \notin B\}$.

EXERCISE 1.4.4 (WITH PARTIAL SOLUTION)

(a) Draw a Venn diagram for difference.

(b) Give an example to show that sometimes $A \backslash B \neq B \backslash A$.

(c) Show (i) $A \backslash A = \varnothing$, (ii) $A \backslash \varnothing = A$.

(d) Show that (i) when $A \subseteq A'$ then $A \backslash B \subseteq A' \backslash B$ and (ii) when $B \subseteq B'$ then $A \backslash B' \subseteq A \backslash B$.

(e) Show that (i) $A \backslash (B \cup C) = (A \backslash B) \cap (A \backslash C)$, (ii) $A \backslash (B \cap C) = (A \backslash B) \cup (A \backslash C)$, and (iii) find a counterexample to $A \backslash (B \backslash C) = (A \backslash B) \backslash C$.

Sample solution to (e)(iii): As a counterexample to (e)(iii), take A to be any non-empty set, e.g. $\{1\}$, and put $C = B = A$. Then LHS $= A\backslash(A\backslash A) = A\backslash\varnothing = A$ while RHS $= (A\backslash A)\backslash A = \varnothing\backslash A = \varnothing$.

The notion of difference acquires particular importance in a special context. Suppose that we are carrying out an investigation into some fairly large set, such as the set $\mathbf{N}^+$ of all positive integers, and that for the purposes of the investigation, the only sets that we need to consider are the subsets of this fixed set. Then it is customary to refer to the large set as a *local universe*, writing it as U, and consider the differences $U \backslash B$ for subsets $B \subseteq U$. As the set U is fixed throughout the investigation, we may as well simplify notation and write $U\backslash B$ alias $U{-}B$ as $-_U B$, or even as simply $-B$ with U left as understood. This application of the difference operation is called *complementation* (within the given universe). Many other notations are also used in the literature for this important operation, e.g. B^-, B', B^c (where the index stands for 'complement'). This time, the Venn diagram needs only one circle, for the set being complemented.

EXERCISE 1.4.5 (WITH PARTIAL SOLUTION)

(a) Draw the Venn diagram for complementation.

(b) Taking the case that A is a local universe U, rewrite equations (e) (i) and (ii) of the preceding exercise using the simple complementation notation described above.

(c) Show that (i) $-(-B) = B$, (ii) $-U = \varnothing$, (iii) $-\varnothing = U$.

Solutions to (b) and (c)(i):

(b) When $A = U$ then equation (i) becomes $U\backslash(B\cup C) = (U\backslash B)\cap(U\backslash C)$, which we can write as $-(B\cup C) = -B\cap -C$; while equation (ii) becomes $U\backslash(B\cap C) = (U\backslash B)\cup(U\backslash C)$, which we can write as $-(B\cap C) = -B\cup -C$.

(c) (i) We need to show that $-(-B) = B$, ie. that $U-(U-B) = B$ whenever $B \subseteq U$ (as assumed when U is taken to be a local universe). We show the two inclusions separately. First, to show $U\backslash(U\backslash B) \subseteq B$, suppose $x \in$ LHS. Then $x \in U$ and $x \notin (U\backslash B)$. From the latter, either $x \notin U$ or $x \in B$, so using the former we have $x \in B =$ RHS as desired. For the converse, suppose $x \in B$. Then $x \notin (U\backslash B)$. But by assumption $B \subseteq U$ so that $x \in U$, and thus $x \in U\backslash(U\backslash B) =$ LHS as desired.

Comments: The identities $-(B\cap C) = -B\cup -C$ and $-(B\cup C) = -B\cap -C$ are known as *de Morgan's laws*, after the nineteenth century mathematician who drew attention to them.

The identity $-(-B) = B$ is known as *double complementation*. Note how its proof made essential use of the hypothesis that $B \subseteq U$.

Logic Box: Negation

You will have noticed that in our discussion of difference and complementation there were a lot of *nots*. In other words, we made free use of the logical connective of negation in our reasoning. What is its logic? Like conjunction and disjunction, it has a truth-table, which is a simple flip-flop:

α	$\neg\alpha$
1	0
0	1

The properties of difference and complementation stem, in effect, from the behaviour of negation used in defining them.

Alice Box: Relative versus absolute complementation

Alice: There is something about this that I don't quite understand. As you define it, the complement $-B$ of a set B is always taken with respect to a given local universe U; it is $U{-}B$. But why not define it in absolute terms? Simply put the absolute complement of B to be the set of all x that are not in B. In other words, take your U to be the set of everything whatsoever.

Hatter: A natural idea indeed – and this is more or less how things were understood in the early days of set theory. Unfortunately it leads to unsuspected difficulties, indeed to contradiction, as was notoriously shown by Bertrand Russell at the beginning of the twentieth century.

Alice: What then?

Hatter: To avoid such contradictions, the standard approach as we know it today, called Zermelo-Fraenkel set theory, does not admit the existence of a universal set, i.e. one containing as elements everything whatsoever. Nor a set of all sets. Nor does it admit the existence of the absolute complement of any set, containing as elements all those things that are not elements of a given set. For if it did, by union it would also have to admit the universal set.

Alice: Is that the only kind of set theory?

Hatter: There are some other versions that do admit absolute complementation and the universal set, for example a system due to Quine. But to avoid

(Continued)

Alice Box: (Continued)

contradiction they must lose power in other respects; and they have other features that tend to annoy working mathematicians and computer scientists. For these reasons they are little used.

Alice: So in this book we are following the standard Zermelo-Fraenkel version?

Hatter: Yes. In practice, the loss of the universal set and absolute complementation are not really troublesome. Whenever you feel that you need to have them, look for a non-universal set that is sufficiently large to contain as elements all the items that you are currently working on, and use it as your local universe for relative complementation.

1.5 Generalised Union and Intersection

It is time to go a little beyond the cosy world of 'flat' set theory, and look at some constructions in which the elements of a set are themselves sets. We begin with the operations of generalised union and intersection.

We know that when A_1, A_2 are sets then we can form their union $A_1 \cup A_2$, whose elements are just those items that are in at least one of A_1, A_2. Evidently, we can repeat the operation, taking the union of that with another set A_3. This will give us $(A_1 \cup A_2) \cup A_3$, and we know from an exercise that this is independent of the order of assembly, i.e. $(A_1 \cup A_2) \cup A_3 = A_1 \cup (A_2 \cup A_3)$, and that its elements are just those items that are elements of at least one of the three. So we might as well write it without brackets.

Clearly we can do this any finite number of times, and so it is natural to consider doing it infinitely many times. In other words, if we have sets A_1, A_2, A_3,...we would like to consider a set $A_1 \cup A_2 \cup A_3 \cup \ldots$ whose elements are just those items in at least one of the A_i for $i \in \mathbf{N}^+$. To make the notation more explicit, we write this set as $\bigcup \{A_i : i \in \mathbf{N}^+\}$ or more compactly as $\bigcup \{A_i\}_{i \in \mathbf{N}+}$ or as $\bigcup_{i \in \mathbf{N}+} \{A_i\}$.

Quite generally, if we have a collection $\{A_i : i \in I\}$ of sets A_i, one for each element i of a fixed set I, we may consider the following two sets :

- $\bigcup_{i \in I} \{A_i\}$, whose elements are just those things that are elements of *at least one* of the A_i for $i \in I$. It is called the *union* of the sets A_i for $i \in I$.

- $\bigcap_{i \in I} \{A_i\}$, whose elements are just those things that are elements of *all* of the A_i for $i \in I$. It is called the *meet* (or *intersection*) of the sets A_i for $i \in I$.

The properties of these general (alias infinite) unions and intersections are similar to those of two-place operations. For example, we have de Morgan and distribution principles. These are the subject of the next exercise.

Alice Box : Sets, collections, familes, classes

Alice: Why do you refer to $\{A_i : i \in I\}$ as as a *collection*, while its elements A_i, and also its union $\bigcup_{i \in I}\{A_i\}$ and its intersection $\bigcap_{i \in I}\{A_i\}$, are called *sets* ?

Hatter: The difference of words does not mark a difference of content. It is merely to make reading easier. The human mind has difficulty in processing phrases like 'set of sets', and even more difficulty with 'set of sets of sets...', and the use of the word 'collection' helps keep us on track.

Alice: I think I have also seen the word 'family'.

Hatter: Here we should be a little careful. Sometimes the term 'family' is used rather loosely to refer to a set of sets. But strictly speaking, it is something different, a certain kind of function. So better not to use that term until it is explained in chapter 3.

Alice: And 'class'?

Hatter: Back at the beginning of the twentieth century, this was used as a synonym for 'set', by people such as Bertrand Russell. And some philosophers continue to use it in that way. But in mathematics, it has acquired a rather special technical sense, beyond the scope of this book. Roughly speaking, a class is something like a set but so large that it cannot be admitted into set theory without contradiction.

Alice: I'm afraid that I don't understand.

Hatter: Don't worry. There are no classes in this book. All you need to remember is that at present you are dealing with *sets*; that sets of sets are also called *collections*; and that families will be introduced in chapter 3.

EXERCISE 1.5.1 (WITH PARTIAL SOLUTION)

From the definitions of general union and intersection, prove the following distribution and de Morgan principles. In the last two, complementation is understood to be relative to an arbitrary sufficiently large universe.

(a) $A \cap \bigcup_{i \in I}\{B_i\} = \bigcup_{i \in I}(A \cap B_i\}$ (distribution of intersection over general union)

(b) $A \cup \bigcap_{i \in I}\{B_i\} = \bigcap_{i \in I}(A \cup B_i\}$ (distribution of union over general intersection)

(c) $-\bigcup_{i\in I}\{A_i\} = \bigcap_{i\in I}\{-A_i\}$ (de Morgan)

(d) $-\bigcap_{i\in I}\{A_i\} = \bigcup_{i\in I}\{-A_i\}$ (de Morgan)

Solution to LHS $\subseteq$ RHS part of (a): This is a simple 'unpack, rearrange, repack' verification. Suppose $x \in$ LHS. Then $x \in A$ and $x \in \bigcup_{i\in I}\{B_i\}$. From the latter, by the definition of general union, we know that $x \in B_i$ for some $i \in I$. So for this $i \in I$ we have $x \in A \cap B_i$, and thus again by the definition of general union, $x \in$ RHS as desired.

1.6 Power Sets

Our next construction is a little more challenging. Let A be any set. We may form a new set, called *the power set* of A, written as $\mathcal{P}(A)$ or 2^A, consisting of all (and only) the subsets of A. In other words, $\mathcal{P}(A) = \{B : B \subseteq A\}$. This may seem like a rather exotic construction, but we will need it as early as chapter 2 when working with relations.

EXERCISE 1.6.1 (WITH SOLUTION)

Let $A = \{1,2,3\}$. List all the elements of $\mathcal{P}(A)$. Using the list, define $\mathcal{P}(A)$ itself by enumeration. How many elements does $\mathcal{P}(A)$ have?

Solution: The elements of $\mathcal{P}(A)$ are (beginning with the smallest and working our way up): $\varnothing$, $\{1\}$, $\{2\}$, $\{3\}$, $\{1,2\}$, $\{1,3\}$, $\{2,3\}$, $\{1,2,3\}$. Thus $\mathcal{P}(A) = \{\varnothing,\{1\},\{2\},\{3\},\{1,2\},\{1,3\},\{2,3\}, \{1,2,3\}\}$. Counting, we see that $\mathcal{P}(A)$ has 8 elements.

Comments: Do not forget the two 'extreme' elements of $\mathcal{P}(A)$: the smallest subset A, namely $\varnothing$, and the largest one, namely A itself. Be careful with the curly brackets. Thus 1, 2, 3 are *not* elements of $\mathcal{P}(A)$, but their singletons $\{1\}$, $\{2\}$, $\{3\}$ are. When defining $\mathcal{P}(A)$ by enumeration, don't forget the outer brackets enclosing the entire list of elements. These may seem like pedantic points of punctuation, but if they are missed then you can get into a dreadful mess.

All of the elements of $\mathcal{P}(A)$ are subsets of A, but some of them are also subsets of others. For example, the empty set is a subset of all of them. This may be brought out clearly by the following *Hasse diagram*, so called after the mathematician who introduced it.

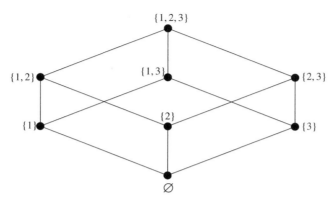

Figure 1.4   *Hasse diagram for P(A) when A = { 1, 2, 3}.*

EXERCISE 1.6.2 (WITH PARTIAL SOLUTION)

Draw Hasse diagrams for the power sets of each of $\varnothing$, $\{1\}$, $\{1,2\}$, $\{1,2,3\}$, $\{1,2,3,4\}$. How many elements does each of these power sets have?

Partial solution: They have 1, 2, 4, 8 and 16 elements respectively.

There is a pattern here. Quite generally, if A is a finite set with n elements, its power set $\mathcal{P}(A)$ has 2^n elements. Here is a rough but very simple proof. Let $a_1,\ldots,a_n$ be the elements of A. Consider any subset $B \subseteq A$. For each a_i there are two possibilities: either $a_i \in B$ or $a_i \notin B$. That gives us $2.2.\ldots.2$ (n times), i.e. 2^n independent choices to determine which among $a_1,\ldots,a_n$ are elements of B, i.e. 2^n possible identities for B.

This fact is very important for computing. Suppose that we have a way of measuring the cost of a computation run in terms of, say, time of calculation as a function of, say, the number of input items. It can happen that this measure increases in proportion to 2^n, i.e. is of the form $k.2^n$ for some fixed k. This is known as *exponential growth*, and it is to be avoided like the plague as it quickly leads to unfeasibly expensive calculations. For example, suppose that such a process is dealing with an input of 10 items. Now $2^{10} = 1024$, which may seem reasonable. But if the input has 100 items to deal with, we have 2^{100} steps to be completed, which would take a very long time indeed.

Logic Box: Truth-table for if

In this chapter we have made frequent use of *if...then...* (alias *conditional*) statements and also *iff* (alias *biconditional*) ones. It is time to fulfill a promise to make their meaning clear. In mathematics, they are used in a very simple way. Like conjunction, disjunction, and negation, they are both truth-functional. The truth-table for the conditional is as follows:

α	β	$\alpha \rightarrow \beta$
1	1	1
1	0	0
0	1	1
0	0	1

From this it is clear that a statement $\alpha \rightarrow \beta$ is always true except when we have the 'disastrous combination': α true and β false.

The biconditional α *iff* β (in longhand, α *if and only if* β, written $\alpha \leftrightarrow \beta$) is less easily true and more easily false. Its table is:

α	β	$\alpha \leftrightarrow \beta$
1	1	1
1	0	0
0	1	0
0	0	1

Thus $\alpha \leftrightarrow \beta$ is true whenever α and β have the same truth-value, and is false whenever they have opposite truth-values. Comparing these tables, it is clear that $\alpha \leftrightarrow \beta$ is true whenever $\alpha \rightarrow \beta$ and its converse $\beta \rightarrow \alpha$ are both true.

Alice Box: The truth-table for the conditional

Alice: Well, at least you kept your promise! But I am not entirely satisfied! I see why the 'disastrous combination' makes *if α then β* false. But why do the other combinations all make it come out true? The two statements α and β may have nothing to do with each other, like 'London is in France' and 'kangaroos are fish'. These are both false, but it is strange to say that the statement 'if London is in France then kangaroos are fish' is true.

(Continued)

Alice Box: (Continued)

Hatter: Indeed, it *is* rather strange and, to be frank, in everyday life we use *if...then...* in very subtle ways – much more complex than any such truth-table. But in mathematics, we use the conditional in the simplest possible manner which, moreover, turns out to underlie all the more complex ones. Here is an example that may not entirely convince you, but might at least make the truth-table less strange. You would agree, I hope, that all positive integers divisible by four are even.

Alice: Of course.

Hatter: Another way of saying the same thing is this: *for every positive integer n, if n is divisible by four then it is even.*

Alice: Indeed.

Hatter: So for every choice that we make of a particular positive integer *n*, the statement *if n is divisible by four then it is even*, comes out true. For example, the following three are all true:

$$If\ 8\ is\ divisible\ by\ 4\ then\ it\ is\ even$$
$$If\ 2\ is\ divisible\ by\ 4\ then\ it\ is\ even$$
$$If\ 9\ is\ divisible\ by\ 4\ then\ it\ is\ even.$$

But the first of these corresponds to the top row of our truth-table (both components true), the second corresponds to the third row (α false, β true), and the last corresponds to the fourth row (both components false), while in each of these three cases the conditional is true.

Alice: I'll have to think about this...

Hatter: ...and think about the following example of Dov Gabbay. A shop hangs a sign in the window saying 'if you buy a computer, then you get a free printer'. The relevant government inspectorate suspects the shop of false advertising, and sends agents disguised as customers into the shop to make purchases. How many ways are there of pinning a false advertising charge on the manager?

1.7 Some Important Sets of Numbers

In this chapter we already came across the set $\mathbf{N}^+ = \{1,2,3,...\}$ of all positive integers. Some other number sets that we will frequently need to refer to, in examples and in general theory, are the following.

$\mathbf{N} = \mathbf{N}^{+} \cup \{0\}$, i.e. the set consisting of zero and all the positive integers. This is called the set of the *natural numbers*. Warning: Some authors use the same term to refer to the positive integers only. Be wary when you read.

$\mathbf{Z} = \{0, \pm 1, \pm 2, \pm 3, \ldots\}$, the set of all *integers* (positive, negative and zero).

$\mathbf{Z}^{-} = \{\ldots, -3, -2, -1\} = \mathbf{Z} \backslash \mathbf{N}$, the set of all *negative integers*.

$\mathbf{Q} = \{p/q : p, q \in \mathbf{Z} \text{ and } q \neq 0\}$, the set of all *rational numbers* (positive, negative and zero).

$\mathbf{R}$ = the set of all *real numbers*, also representable as the set of all numbers of the form $p + d_1 d_2 \ldots$ where $p \in \mathbf{Z}$ and $d_1 d_2 \ldots$ is an ending or unending decimal (series of digits from 0 to 9).

We will often need to refer to these sets, and assume at least a little familarity with them, especially the first five.

FURTHER EXERCISES

1.1. *Boolean operations on sets*

We define the operation $A + B$ of *symmetric difference* (sometimes known as disjoint union) by the following rule: $A + B = (A \backslash B) \cup (B \backslash A)$. Notations vary: often $\oplus$ is used for this operation, sometimes Δ.

(a) Show that for any x, $x \in A + B$ iff x is an element of exactly one of A, B.

(b) Draw a Venn diagram for the operation.

(c) Show that $A + B \subseteq A \cup B$

(d) Show that $A + B$ is disjoint from $A \cap B$

(e) Show that $A + B = (A \cup B) \backslash (A \cap B)$

(f) For each of the following properties of $\cup$, check out whether or not it also holds for $+$, giving a proof or a counterexample as appropriate: (i) commutativity, (ii) associativity, (iii) distribution of $\cap$ over $+$, (iv) distribution of $+$ over $\cap$.

(g) Express $-(A + B)$ using union, intersection, complement.

(h) We have seen that each of intersection, union and difference corresponds to a truth-functional logical connective. To what connective does symmetric difference correspond? Draw its truth-table. *Hint*: Read again the Logic Box on disjunction.

1.2. *Counting principles for union and intersection*

Let A, B be finite sets. We write $\#(A)$, $\#(B)$ for the number of elements that each contains.

(a) Show that always $\#(A \cup B) \leq \#(A) + \#(B)$.

(b) Give an example to show that sometimes $\#(A \cup B) < \#(A) + \#(B)$.

(c) Show that when A, B are disjoint then $\#(A \cup B) = \#(A) + \#(B)$.

(d) Show quite generally that when A, B are any finite sets (not necessarily disjoint), then $\#(A \cup B) = \#(A) + \#(B) - \#(A \cap B)$. This is known as the *rule of inclusion and exclusion*. Despite its simplicity, it will be one of the two fundamental principles of the chapter on combinatorics.

(e) Show that $\#(A \cap B) \leq min(\#(A), \#(B))$. Here, $min(m,n)$ is whichever is the lesser of the integers m,n.

(f) Give an example to show that sometimes $\#(A \cap B) < min(\#(A), \#(B))$.

(g) Formulate and verify a necessary and sufficient condition for the equality $\#(A \cap B) = min(\#(A), \#(B))$ to hold.

1.3. *Counting principles for complements*

(a) Using the same notation as in the preceding exercise, show that $\#(A \backslash B) = \#(A) - \#(A \cap B)$.

(b) Use this to show that when $B \subseteq A$ then $\#(A \backslash B) = \#(A) - \#(B)$.

(c) Give an example showing that the equality in (b) may fail when we drop the condition $B \subseteq A$.

1.4 *Generalized union and intersection*

(a) Let $\{A_i\}_{i \in I}$ be any collection of sets. Show that for any set B we have (i) $\bigcup \{A_i\}_{i \in I} \subseteq B$ iff $A_i \subseteq B$ for every $i \in I$, (ii) $B \subseteq \bigcap \{A_i\}_{i \in I}$ iff $B \subseteq A_i$ for every $i \in I$.

(b) Find a collection $\{A_i\}_{i \in I}$ of non-empty sets with each $A_i \supset A_{i+1}$ but with $\bigcap \{A_i\}_{i \in I}$ empty. *Hint*: You might as well look among the subsets of $\mathbf{N}$.

1.5. *Power sets*

(a) Show that whenever $A \subseteq B$ then $\mathcal{P}(A) \subseteq \mathcal{P}(B)$.

(b) True or false? $\mathcal{P}(A \cap B) = \mathcal{P}(A) \cap \mathcal{P}(B)$. If true, prove it; if false, give a counterexample.

(c) True or false? $\mathcal{P}(A \cup B) = \mathcal{P}(A) \cup \mathcal{P}(B)$. If true, prove it; if false, give a counterexample.

Selected Reading

A classic of beautiful eposition, but short on exercises:

Paul R. Halmos *Naive Set Theory*. Springer, 2001 (new edition), Chapters 1–5, 9.

The present material is covered with lots of exercises in:

Seymour Lipschutz *Set Theory and Related Topics*. McGraw Hill Schaum's Outline Series, 1998, Chapters 1–2 and 5.1–5.3.

A recent presentation written for first year mathematics students:

Carol Schumacher *Chapter Zero*: *Fundamental Notions of Abstract Mathematics*. Pearson, 2001 (second edition), Chapter 2.

Comparing Things: Relations

Chapter Outline

Relations play an important role in computer science, both as tools of analysis and as instruments for representing computational structures such as databases. In this chapter we introduce the basic concepts you need to master in order to work with them.

We begin with the notions of an *ordered pair* (and more generally, ordered *n-tuple*) and the *Cartesian product* of two more or more sets. We then consider operations on relations, notably those of forming the *converse*, *join*, and *composition* of relations, as well as some other operations that combine both relations and sets, notably those of the *image* and the *closure* of a set under a relation.

We also explore two of the main jobs that relations are asked to carry out: to classify and to order. For the former, we explain the notion of an *equivalence relation* (reflexive, transitive, symmetric) over a set and how it corresponds to the notion of a *partition* of the set. For the latter, we look first of all at several kinds of *reflexive order*, and then at their *strict parts*.

2.1 Ordered Tuples, Cartesian Products, Relations

What do the following have in common? One car overtaking another, a boy loving a girl, one tree being shadier than another, an integer dividing another, a point lying between two others, and a student exchanging one book for another with a friend.

D. Makinson, *Sets, Logic and Maths for Computing*,
DOI: 10.1007/978-1-84628-845-6_2, © Springer-Verlag London Limited 2008

They are all examples of *relations* involving at least two items – in some instances three (one point between two others), four (the book exchange), or more. Often they involve actions, intentions, the passage of time, and causal connections; but in mathematics and computer science we abstract from all those features and work with a very basic, stripped-down concept. To explain what it is, we begin with the notions of an ordered tuple and Cartesian product.

2.1.1 Ordered Tuples

Recall from the preceding chapter that when a set has exactly one element, it is called a *singleton*. When it has exactly two distinct elements, it is called a *pair*. For example, the set $\{7,9\}$ is a pair, and it is the same as the pair $\{9,7\}$. We have $\{7,9\} = \{9,7\}$ because the order is irrelevant: the two sets have exactly the same elements.

An *ordered pair* is like a (plain, unordered) pair except that order matters. To highlight this, we use a different notation. The ordered pair whose first element is 7 and whose second element is 9 is written as $(7,9)$ or, in older texts, as $<7,9>$. It is distinct from the ordered pair $(9,7)$ although they have exactly the same elements: $(7,9) \neq (9,7)$, because the elements are considered in a different order.

Abstracting from this example, the *criterion for identity* of ordered pairs is as follows: $(x_1,x_2) = (y_1,y_2)$ iff both $x_1 = y_1$ and $x_2 = y_2$. This contrasts with the criterion for identity of plain sets: $\{x_1,x_2\} = \{y_1,y_2\}$ iff the left and right hand sets have exactly the same elements, which (it is not difficult to show) holds iff *either* $(x_1 = y_1$ and $x_2 = y_2)$ *or* $(x_1 = y_2$ and $x_2 = y_1)$.

More generally, the criterion for identity of two ordered n-tuples $(x_1,x_2,\ldots,x_n)$ and $(y_1,y_2,\ldots,y_n)$ is as you would expect: $(x_1,x_2,\ldots,x_n) = (y_1,y_2,\ldots,y_n)$ iff $x_i = y_i$ for all i from 1 to n.

Alice Box: Ordered pairs

Alice: I have a technical problem here. Aren't there other ways in which plain pairs can be identical? For example, when $x_1 = y_1$ and $x_2 = y_1$ and $y_2 = x_1$ then the two sets $\{x_1,x_2\}$ and $\{y_1,y_2\}$ have exactly the same elements, and so are identical.

Hatter: Sure they are. But then, since $x_1 = y_1$ and $x_2 = y_1$ we have $x_1 = x_2$, so since $y_2 = x_1$ we have $x_2 = y_2$. Thus $x_1 = y_1$ and $x_2 = y_2$, so that the situation that you are considering is already covered by the first of the two cases that were mentioned in the definition.

(Continued)

Alice Box: (Continued)

Alice: OK. But I also have a philosophical problem. Isn't there something circular in all this? You promised that relations will be used to build a theory of order, but here you are defining the concept of a relation by using the notion of an ordered pair, which already involves the concept of order!

Hatter: A subtle point, and a good one! But I would call it a spiral rather than a circle. We need just a *rock-bottom* kind of order – no more than the idea of one thing coming before another – in order to understand what an ordered pair is. From that we can build a very sophisticated theory of the various kinds of order that relations can create.

EXERCISE 2.1.1

Check in detail the Hatter's claim that $\{x_1, x_2\} = \{y_1, y_2\}$ iff either ($x_1 = y_1$ and $x_2 = y_2$) or ($x_1 = y_2$ and $x_2 = y_1$). *Hint*: You may find it helps to break your argument into cases.

2.1.2 Cartesian Products

With this in hand, we can introduce the notion of the Cartesian product of two sets. If A, B are sets then their *Cartesian product*, written $A \times B$ and pronounced 'A cross B' or 'A by B', is defined as follows:

$$A \times B = \{(a, b) : a \in A \text{ and } b \in B\}$$

In English, $A \times B$ is the set of all ordered pairs whose first term is in A and whose second term is in B. When $B = A$, so that $A \times B = A \times A$ it is customary to write it as A^2, calling it 'A squared'.

A very simple concept, but be careful – it is also easy to trip up! Take note of the *and* in the definition, but be careful not to confuse Cartesian products with intersections. For example, if A, B are sets of numbers, then $A \cap B$ is also a set of *numbers*; but $A \times B$ is a set of *ordered pairs of numbers*.

EXERCISE 2.1.2 (WITH SOLUTION)

Let $A = \{\text{John, Mary}\}$ and $B = \{1, 2, 3\}$, $C = \varnothing$. What are $A \times B$, $B \times A$, $A \times C$, $C \times A$, A^2, B^2? Are any of these identical with others? How many elements in each?

Solution:

$A \times B = \{(\text{John},1), (\text{John},2), (\text{John},3), (\text{Mary},1), (\text{Mary},2), (\text{Mary},3)\}$

$B \times A = \{(1,\text{John}), (2,\text{John}), (3,\text{John}), (1,\text{Mary}), (2,\text{Mary}), (3,\text{Mary})\}$

$A \times C = \varnothing$

$C \times A = \varnothing$

$A^2 = \{(\text{John},\text{John}), (\text{John, Mary}), (\text{Mary},\text{John}), (\text{Mary, Mary})\}$

$B^2 = \{(1,1), (1, 2), (1,3), (2,1), (2, 2), (2,3), (3,1), (3, 2), (3,3)\}$

Of these, $A \times C = C \times A$, but that is all; in particular $A \times B \neq B \times A$. Counting the elements: $\#(A \times B) = 6 = \#(B \times A)$, $\#(A \times C) = 0 = \#(C \times A)$, $\#(A^2) = 4$, $\#(B^2) = 9$.

Comment: Note that also $A \times B = \{(\text{John},1), (\text{Mary},1), (\text{John},2), (\text{Mary},2), (\text{John},3), (\text{Mary},3)\}$. Within the curly brackets we are enumerating the elements of a *set*, so we can write them in any order we like; but within the round brackets we are enumerating the terms of an *ordered pair* (or ordered n-tuple), so there the order is vital.

The operation takes its name from René Descartes who, in the seventeenth century, made use of the Cartesian product $\mathbf{R}^2$ of the set $\mathbf{R}$ of all real numbers. His seminal idea was to represent each point of a plane by an ordered pair (x,y) of real numbers, and use this representation to solve geometric problems by algebraic methods. The set $\mathbf{R}^2$ is called the *Cartesian plane*.

From the exercise, you may already have guessed a general counting principle for the Cartesian products of finite sets: $\#(A \times B) = \#(A) \cdot \#(B)$ where the dot stands for ordinary multiplication. Here is a rough proof. Let $\#(A) = m$ and $\#(B) = n$. Fix any element $a \in A$. Then there are n different pairs (a,b) with $b \in B$. And when we fix a different $a' \in A$, then the n pairs (a',b) will all be different from the pairs (a,b), since they differ on their first terms. Thus there are $n + n + \ldots + n$ (m times), i.e. $m \cdot n$ pairs altogether in $A \times B$.

Thus although the operation of forming the Cartesian product of two sets is *not* commutative (i.e. we may have $A \times B \neq B \times A$), the operation of counting the elements of the Cartesian product *is* commutative, i.e. always $\#(A \times B) = \#(B \times A)$.

EXERCISE 2.1.3 (WITH PARTIAL SOLUTION)

(a) Show that when $A \subseteq A'$ and $B \subseteq B'$ then $A \times B \subseteq A' \times B'$.

(b) Show that when both $A \neq \varnothing$ and $B \neq \varnothing$ then $A \times B = B \times A$ iff $A = B$

Solution to (b): Suppose $A \neq \varnothing$ and $B \neq \varnothing$. We need to show that $A \times B = B \times A$ iff $A = B$. We do this in two parts.

First, we show that if $A = B$ then $A \times B = B \times A$. Suppose $A = B$; we need to show $A \times B = B \times A$. By the supposition, $A \times B = A^2$ and also $B \times A = A^2$ so that $A \times B = B \times A$ as desired.

Next, we show the converse, that if $A \times B = B \times A$ then $A = B$. The easiest way to do this is by showing the *contrapositive*: if $A \neq B$ then $A \times B \neq B \times A$. Suppose $A \neq B$. Then either $A \nsubseteq B$ or $B \nsubseteq A$. We consider the former case; the latter is similar. Since $A \nsubseteq B$ there is an $a \in A$ with $a \notin B$. By supposition, $B \neq \varnothing$, so there is a $b \in B$. Thus $(a,b) \in A \times B$ but since $a \notin B$, $(a,b) \notin B \times A$. Thus $A \times B \nsubseteq B \times A$ as desired.

Logic Box: Proof of 'iff' statements, and proof by contraposition

The solution to Exercise 2.1.3(b) is instructive in several respects. It uses a 'divide and rule' strategy, breaking the problem down into component parts, and tackling them patiently one by one. Within each part (or sub-problem), we make an appropriate supposition and work to the corresponding goal. In this way, a quite complex problem can often be reduced to a collection of very simple ones.

In particular, in order to prove an *if and only if* statement, it is often convenient to do it in two parts: first prove the *if* in one direction, and then prove it in the other. As we saw in the preceding chapter, these are not the same; they are called *converses* of each other.

Finally, the example also illustrates the process of proving by contraposition. Suppose we want to prove a statement *if α then β*. As mentioned in Chapter 1, the most straightforward way of tackling this is to suppose α and drive towards β. But that does not always give the most transparent proof. Sometimes it is better to suppose *not-β* and head for *not-α*. That is what we did in the example: to prove that if $A \times B = B \times A$ then $A = B$ for nonempty sets A, B, we supposed $A \neq B$ and showed $A \times B \neq B \times A$.

Why is this method of proof legitimate? Because the two conditionals $\alpha \rightarrow \beta$ and $\neg\beta \rightarrow \neg\alpha$ are equivalent, as can be seen by examining their truth-tables.

How can it help? Often, the supposition $A \neq B$ gives us something to 'grab hold of'. In our example, it tells us that there is an a with $a \in A$, $a \notin B$ (or conversely); we can then consider a particular such a, and start reasoning about it. This pattern occurs quite often.

2.1.3 Relations

Let A, B be any sets. A *binary relation from A to B* is defined to be any subset of
the Cartesian product $A \times B$. It is thus any set of ordered pairs (a, b) such $a \in A$ and
$b \in B$. A binary relation is therefore fully determined by the ordered pairs that it
covers. It does not matter how these pairs are presented or described. It is
customary to use R, S, ... as symbols standing for relations. As well as saying
that the relation is 'from A to B', one also says that it is '*over $A \times B$*'.

From the definition, it follows that in the case that $A = B$, a binary relation
from A to A is any set of ordered pairs (a, b) such both $a, b \in A$. It is thus a relation
over A^2, but informally we often abuse language a little and describe it as a
relation *over A*.

Evidently, the notion may be generalised to any number of places. Let
$A_1, \ldots, A_n$ be sets. An *n-place relation over $A_1 \times \ldots \times A_n$* is defined to be any subset
of $A_1 \times \ldots \times A_n$. In other words, it is any set of *n-tuples* $(a_1, \ldots, a_n)$ with each $a_i \in A_i$.

In this chapter we will be concerned mainly with binary relations, and when
there is no ambiguity will speak of them simply as relations.

EXERCISE 2.1.4 (WITH SOLUTION)

Let A, B be as in Exercise 2.1.2.

(a) Which of the following are (binary) relations from A to B?

    (i) {(John,1), (John,2)}

    (ii) {(Mary,3), (John,Mary)}

    (iii) {(Mary,2), (2,Mary)}

    (iv) {(John,3), (Mary,4)}

    (v) ({Mary,1}, {John,3})

(b) What is the largest relation from A to B? What is the smallest?

(c) Identify (by enumeration) three more relations from A to B.

(d) How many relations are there from A to B?

Solution:

(a) Only (i) is a relation from A to B. (ii) is not, because Mary is not in B.
(iii) is not, because 2 is not in A. (iv) is not, because 4 is not in B. (v) is
not, because it is a an ordered pair of sets, not a set of ordered pairs.

(b) $A \times B$ is the largest, $\varnothing$ is the smallest, in the sense that $\varnothing \subseteq R \subseteq A \times B$ for every relation R from A to B.

(c) For brevity, we can choose three *singleton relations*: $\{(\text{John},1)\}$, $\{(\text{John},2)\}$, $\{(\text{John},3)\}$.

(d) Since $\#(A) = 2$ and $\#(B) = 3$, $\#(A \times B) = 2 \bullet 3 = 6$. From the definition of a relation from A to B it follows that the set of all relations from A to B is just $\mathcal{P}(A \times B)$, and by a principle in the chapter on sets, $\#(\mathcal{P}(A \times B)) = 2^{\#(A \times B)} = 2^6 = 64$.

When R is a relation from A to B, we call the set A a *source* of the relation, and B a *target*. Sounds simple enough, but care is advised. As already noted in an exercise, when $A \subseteq A'$ and $B \subseteq B'$ then $A \times B \subseteq A' \times B'$, so when R is a relation from A to B then it is also a relation from A' to B'. Thus the source and target of R, in the above sense, are not unique: a single relation will have indefinitely many sources and targets.

For this reason, we also need terms for the least possible source and target of R. We define the *domain* of R to be the set of all a such that $(a,b) \in R$ for some b, writing briefly $dom(R) = \{a$: there is a b with $(a,b) \in R\}$. Likewise we define $range(R) = \{b$: there is an a with $(a,b) \in R\}$. Clearly, whenever R is a relation from A to B then $dom(R) \subseteq A$ and $range(R) \subseteq B$.

Warning: You may occasionally see the term 'codomain' contrasting with 'domain'. But care is needed, as the term is sometimes used broadly for 'target', sometimes more specifically for 'range'. In this book we will follow a fairly standard terminology, with *domain* and *range* defined as above, and *source*, *target* for any supersets of them.

EXERCISE 2.1.5

(a) Consider the relation $R = \{(1,7), (3,3), (13,11)\}$ and the relation $S = \{(1,1), (3,11), (13,12), (15,1)\}$. Identify $dom(R)$, $range(R)$, $dom(S)$, $range(S)$.

(b) The *identity relation* I_A over a set A is defined by putting $I_A = \{(a,a) : a \in A\}$. Identify $dom(I_A)$ and $range(I_A)$.

(c) Identify $dom(A \times B)$, $range(A \times B)$.

Since relations are sets (of ordered pairs or tuples), we can apply to them all the concepts that we developed for sets. In particular, it makes sense to speak of one relation R being included in another relation S: every tuple that is an element of R is an element of S. In this case we also say that R is a *subrelation* of S.

Likewise, it makes sense to speak of the *empty relation*: it is the relation that has no elements, and it is unique. It is thus the same as the empty set, and can be written ∅.

EXERCISE 2.1.6

    (a) Use the definitions to show that (i) the empty relation is a subrelation of every relation, (ii) the empty relation has no proper subrelations.

    (b) Identify $dom(\emptyset)$, $range(\emptyset)$.

    (c) What would it mean to say that two relations are disjoint? Give an example of two disjoint relations over a small finite set A.

2.2 Tables and Digraphs for Relations

In mathematics, rigour is important, but so is intuition. The two should go hand in hand. One way of strengthening one's intuition is to use graphic representations. This is particularly so in the case of binary relations. For sets in general we used Euler and Venn diagrams; for the more specific case of relations, tables and arrow diagrams are helpful.

2.2.1 Tables for Relations

Let's go back to the sets $A = \{$John, Mary$\}$ and $B = \{1,2,3\}$ of earlier exercises. Consider the relation $R = \{$(John,1), (John,3), (Mary,2), (Mary,3)$\}$. How might we represent R by a table?

    We draw a table with two rows and three columns, for the elements of A and B respectively. Each cell in this table is uniquely identified by its coordinate (a,b) where a is the element for the row and b is the element for the column. Write in the cell a 1 (for 'true') or a 0 (for 'false') according as the ordered pair (a,b) is or is not an element of the relation. For the R chosen above, this gives us the following table.

Table 2.1 Table for a relation.

R	1	2	3
John	1	0	1
Mary	0	1	1

Tabular representations of relations are particularly useful when dealing with databases, because good software is available for writing and manipulating them. When a table has the same number of columns as rows, geometrical operations such as folding along the diagonal can also reveal interesting structural features.

EXERCISE 2.2.1

(a) Let $A = \{1,2,3,4\}$. Draw tables for each of the following relations over A^2: (i) $<$, (ii) $\leq$, (iii) $=$, (iv) $\varnothing$, (v) A^2.

(b) In each of the four tables, draw a line from the top left cell (1,1) to the bottom right cell (4,4). This is called the *diagonal* of a relation over A^2. Comment on the contents of the diagonal in each of the four tables.

(c) Imagine folding the table along the diagonal (or cut it out and fold it). Comment on any symmetries that become visible.

2.2.2 Digraphs for Relations

Another way of representing binary relations, less suitable for software implementation but friendly to humans, is by means of arrow diagrams known as *directed graphs* or more briefly *digraphs*. The idea is simple, at least in the finite case. Given a relation from A to B, mark a point for each element of $A \cup B$, labelling it with a name if desired. Draw an arrow from one point to another just when the first stands in the relation to the second. When the source A and target B of the relation are not the same, it can be useful to add into the diagram circles for the sets A and B.

In the example considered for the table above, the digraph comes out as follows.

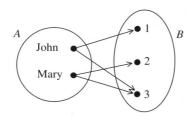

Figure 2.1   *Diagram for a relation.*

The Hasse diagram between the subsets of $\{1,2,3\}$ that we presented in Chapter 1 can be seen as a digraph for the relation of *being immediately included in*. This holds between a set B and a set C iff $B \subset C$ but there is no X with $B \subset X \subset C$. However, the Hasse diagram uses plain lines rather than arrows, with the convention that these are always read with subset below and superset above.

Given a Hasse diagram for a relation, we can use it to read off its reflexive and transitive closure. In particular, given the Hasse diagram for the relation of B being *immediately included* in C, we can read off the relation of B being *included* in C: it holds iff there is an ascending path from B to C (including the one-element path). This is much more economical than drawing a digraph for the entire relation of inclusion.

EXERCISE 2.2.2

Take from Chapter 1 the Hasse diagram for immediate inclusion between subsets of the set $A = \{1,2,3\}$, and compare it with the digraph for the entire relation of inclusion. How many links in each?

Diagrams are valuable tools to illustrate situations and stimulate intuitions. They can often help us think up counterexamples to general claims, and they can sometimes be used to illustrate a proof, making it much easier to follow. However, they have their limitations. In particular, it should be remembered that *a diagram can never itself constitute a proof of a general claim*.

2.3 Operations on Relations

Since relations are sets, we can carry out on them all the Boolean operations for sets that we learned in the preceding chapter, provided we keep track of the sources and targets. Thus, if R and S are relations from the same source A to the same target B, their intersection $R \cap S$, being the set of all ordered pairs (x,y) that are simultaneously elements of R and of S, will also be a relation from A to B. Likewise for the union $R \cup S$, and also for complement $-R$ with respect to $A \times B$. Note that just as for sets, $-R$ is really a difference $(A \times B) - R$ and so depends implicitly on the source A and target B as well as on R itself.

As well as the Boolean operations, there are others that arise only for relations. We describe some of the most important ones.

2.3.1 Converse

The simplest is that of forming the *converse* (alias *inverse*) R^{-1} of a relation. Given a relation R, we define R^{-1} to be the set of all ordered pairs (b,a) such that $(a,b) \in R$.

Warning: Do not confuse this with complementation! There is nothing negative about conversion: we are simply *reversing the direction* of the relation. The complement of *loves* is *doesn't love*, but its converse is *loved by*.

EXERCISE 2.3.1 (WITH SOLUTIONS)

(a) Let A be the set of natural numbers. What are the converses of the following relations over A: (i) less than, (ii) less than or equal to, (iii) equal to.

(b) Let A be a set of people. What are the converses of the following relations over A: (i) being a child of, (ii) being a descendant of, (iii) being a daughter of, (iv) being a brother of, (v) being a sibling of, (vi) being a husband of.

Solutions:

(a) (i) greater than, (ii) greater than or equal to, (iii) equal to.

(b) (i) being a parent of, (ii) being an ancestor of, (iii) having as a daughter, (iv) having as a brother, (v) being a sibling of, (vi) being a wife of.

Comments: In group (a) we already have examples of how the converse of a relation may be disjoint from it, overlap with it, or be identical with it.

Some of the examples in group (b) are a little tricky. Note that the converse of being a brother of is *not* being a brother of: when a is a brother of b, b may be female and so a sister of a.

Sometimes, ordinary language has a single word for a relation and another single word for its converse. This is the case for (i) (child/parent) (ii) (ancestor/descendant) and (iii) (husband/wife). But it is not always so: witness (iii) and (iv) where we have to use special turns of phrase: daughter/having as daughter, brother/having as brother.

EXERCISE 2.3.2

(a) What does the converse of a relation look like from the point of view of a digraph for the relation? And from the point of view of a table for it?

(b) Show that $dom(R^{-1}) = range(R)$ and $range(R^{-1}) = dom(R)$.

(c) Show that (i) $(R^{-1})^{-1} = R$, (ii) $(R \cap S)^{-1} = R^{-1} \cap S^{-1}$, (iii) $(R \cup S)^{-1} = R^{-1} \cup S^{-1}$. Compare these equalities with those for complementation.

Alice Box: Converse of an n-place relation

Alice: Does the notion of conversion make sense for relations with more than two places?

Hatter: It does, but there we have not just one operation but several. For simplicity, take the case of a three-place relation R, whose elements are ordered triples (a,b,c). Then there is an operation that converts the first two, giving triples (b,a,c), and an operation converting the last two, giving triples (a,c,b).

Alice: And one switching the first and the last, giving triples (c,b,a)?

Hatter: Indeed. However once we have some of these operations we can get others by iterating them. For example, your operation may be obtained by using the first two as follows: from (a,b,c) to (b,a,c) to (b,c,a) to (c,b,a), using the first, second, and first operations. We could also get the first operation, say, by iterating the second and third...

Alice: Stop there, I've got the idea.

2.3.2 Join of Relations

Imagine that you are in the personnel division of a firm, and that you are in charge of a database recording the identity numbers and names of employees. Your colleague across the corridor is in charge of another database recording their names and telephone extensions. Each of these databases may be regarded as a binary relation. In effect, we have a large set A of allowable identity numbers (e.g. any six digit figure), a set B of allowable names (e.g. any string of at most twenty letters and spaces, with no space at beginning or end and no space immediately following another one), and a set C of allowable telephone extension numbers (e.g. any figure of exactly four digits). Your database is a subset R of $A \times B$, your colleague's database is a subset S of $B \times C$; they are thus both relations. These relations may have special properties. For example, it may be required that R cannot associate more than one name with any given identity number, although S may legitimately associate more than one telephone extension to a given name. We will not bother with these special properties at the moment, leaving them to the chapter on functions.

The manager may decide to merge these two databases into one, thus economising on the staff needed to maintain them and liberating one of you for other tasks (or for redundancy). What would the natural merging be? A three-place relation over $A \times B \times C$ consisting of all those triples (a,b,c) such that $(a,b) \in R$ and $(b,c) \in S$. This is clearly an operation on relations, taking two binary relations with a common middle set (target of one, source of the other) to form a three-place relation. The resulting relation/database gives us all the data of the two components, with no loss of information.

Of course, in practice, databases make use of relations with more than just two places, and the operation that we have described may evidently be generalised to cover them. Let R be a relation over $A_1 \times \ldots \times A_m \times B_1 \times \ldots \times B_n$ and let S be a relation over $B_1 \times \ldots \times B_n \times C_1 \times \ldots \times C_p$. We define *join* of R and S, written *join*(R,S), to be the set of all those $(m+n+p)$-tuples $(a_1,\ldots,a_m,b_1,\ldots,b_n,c_1,\ldots,c_p)$ such that $(a_1,\ldots,a_m,b_1,\ldots,b_n) \in R$ and $(b_1,\ldots,b_n,c_1,\ldots,c_p) \in S$.

Another terminological warning: database theorists (and this book) call this operation 'join'; set theorists and algebraists sometimes use the same word as a synonym for the simple union of sets, as noted in the preceding chapter.

By combining operations of join and conversion, one may do quite a lot of manipulation. These operations never diminish the *arity* (i.e. number of places) of the relations: conversion leaves the arity unchanged, join increases it. But in database theory there are several further operations that cannot be obtained from conversion and join alone. One is *projection*. Suppose $R \subseteq A_1 \times \ldots \times A_m \times B_1 \times \ldots \times B_n$ is a database. We may *project* R onto its first m places, forming a relation whose elements are just those m-tuples $(a_1,\ldots,a_m)$ such that there are $b_1,\ldots,b_n$ with $(a_1,\ldots,a_m,b_1,\ldots,b_n) \in R$.

Another database operation is *selection* (alias *restriction* in the language of set theorists and algebraists). Again, let $R \subseteq A_1 \times \ldots \times A_m \times B_1 \times \ldots \times B_n$ be a database, and let $C_1,\ldots,C_m$ be subsets of $A_1,\ldots,A_m$ respectively (i.e. $C_i \subseteq A_i$ for each $i \leq m$). We may *select* (or restrict the relation to) the subsets by taking its elements to be just those $(m+n)$-tuples $(c_1,\ldots,c_m,b_1,\ldots,b_n)$ such that $(c_1,\ldots,c_m,b_1,\ldots,b_n) \in R$ and each $c_i \in C_i$.

Here we have selected subsets from the first m arguments. We could equally well have selected from the last n, or from any others. In database contexts, the subsets C_i will often be singletons $\{c_i\}$.

EXERCISE 2.3.3

Formulate the definition of selection (alias restriction) for the special case that we have a two-place relation over $A \times B$ and we restrict (i) just A, (ii) both A and B.

2.3.3 Composition of Relations

While the join operation is immensely useful for database manipulation, it does not occur very often in everyday language. But there is a variant of it that children learn at a very young age – as soon as they can recognise members of their extended family. It was first studied in the nineteenth century by Augustus de Morgan who called it 'relative product'; nowadays it is usually called 'composition'.

Suppose we are given two relations $R \subseteq A \times B$ and $S \subseteq B \times C$, with the target of the first the same as the source of the second. We have already defined their join as the set of all triples (a,b,c) such that $(a,b) \in R$ and $(b,c) \in S$. But we can also define their *composition* $S \circ R$ as the set of all ordered pairs (a,c) such *that there is some x* with both $(a,x) \in R$ and $(x,c) \in S$.

For example, if F is the relation of 'a father of' and P is the relation of 'a parent of', then $P \circ F$ is the relation consisting of all ordered pairs (a,c) such that there is some x with $(a,x) \in F$ and $(x,c) \in P$, i.e. all ordered pairs (a,c) such that for some x, a is a father of x and x is a parent of c. It is thus the relation of being 'a father of a parent of', i.e. 'a grandfather of'.

Alice Box: Notation for composition of relations

Alice: That feels funny, the wrong way round. Wouldn't it be easier to write the composition $P \circ F$ with the letters *the other way round*, so that they follow the same order as they are mentioned in the phrase 'a father of a parent of', which is also the order of occurrence of the predicates in the phrase '$(a,x) \in F$ and $(x,c) \in P$' of the definition?

Hatter: Indeed it would be easier, and most of the earlier authors working in the theory of relations did it the way that you suggest.

Alice: Why the switch?

Hatter: To bring it into agreement with the way we usually do things in the theory of functions. As you will see in the next chapter, a function can be defined as a special kind of relation, and notions such as composition for functions turn out to be the same as for relations; so it is best to use the same notation. In this case, the notation for functions won out. A pity, as I am a relations man myself.

Alice: I'm afraid that I'll always mix them up!

Hatter: The best policy is to commit the definition to memory, and write it down before each exercise involving composition of relations.

Logic Box: Existential quantifier

The definition of the composition of two relations makes essential use of the phrase 'there is an x such that . . .'. This is known as the *existential quantifier*, and is written $\exists x(\ldots)$. For example, 'There is an x with both $(a,x) \in R$ and $(x,c) \in S$ is written as $\exists x((a,x) \in R \wedge (x,c) \in S)$, or more briefly as $\exists x(Rax \wedge Sxc)$.

Spoken language does not contain variables; nor does written language outside mathematical contexts. But it manages to express existential quantification in other ways. For example, if we say 'Some composers are poets' we are not using a variable, but we are in effect saying that there is an x such that x is both a composer and a poet, which the set-theorist would write as $\exists x(x \in C \wedge x \in P)$ and the logician would write as $\exists x(Cx \wedge Px)$. Pronouns can also be used like variables. For example, when we say 'If there is a free place, I will reserve it', the 'it' is doing the work of a variable. However, once we begin to formulate more complex statements involving several quantifications, it can be difficult to be precise without using variables explicitly.

The existential quantifier $\exists$ has its own logic, which we will describe later, along with the logic of its companion, the universal quantifier $\forall$. In the meantime, we will simply adopt the notation $\exists x(\ldots)$ as a convenient shorthand for longer English 'there is an x such that . . .', and likewise $\forall x(\ldots)$ to abbreviate 'for every x, . . . holds'.

EXERCISE 2.3.4 (WITH SOLUTION)

Let A be the set of people, and P, F, M, S, B the relations over A of 'parent of', 'father of', 'mother of', 'sister of' and 'brother of' respectively. Describe *exactly* the following relative products. (a) $P \circ P$, (b) $M \circ F$, (c) $S \circ P$, (d) $B \circ B$. *Warnings*: (1) Be careful about order. (2) In some cases there will be a handy word in English for just the relation, but in others it will have to be described in a more roundabout (but still precise) way.

Solution and remarks:

(a) $P \circ P = $ 'grandparent of'. *Reason*: a is a grandparent of c iff there is an x such that a is a parent of x and x is a parent of c.

(b) $M \circ F = $ 'maternal grandfather of'. *Reason*: a is a maternal grandfather of c iff there is an x such that a is a father of x and x is a mother of c. *Comments*: Two common errors here. (1) Getting the order wrong. (2) Rushing to the answer 'grandfather of', since a father of a mother is always a grandfather. But there is another way of being a grandfather, namely by being a father of a father, so that relation is too broad.

(c) $S \circ P$ = 'parent of a sister of'. *Reason*: a is a parent of a sister of c iff there is an x such that a is a parent of x and x is a sister of c. *Comments*: This one is tricky. When there is an x such that a is a parent of x and x is a sister of c, then a is also a parent of c, which tempts one to rush into the answer 'parent of'. But that relation is also too broad, because a may also be a parent of c when c has no sisters! English has no single word for the relation $S \circ P$; we can do no better than use a circumlocution such as 'parent of a sister of'.

(d) $B \circ B$ = 'brother of a brother of'. *Comments*: Again, English has no single word for the relation. One is tempted to say that $B \circ B$ = 'brother'. But that answer is both too broad (in one respect) and too narrow (in another respect). Too broad because a may be a brother of c without being a brother of a brother x of c: there may be no third brother to serve as the x. Too narrow because a may be a brother of x and x a brother of c, without a being a brother of c, for a may be the same person as c! In this example, again, we can do little better than use the phrase 'brother of a brother of'.

Different languages categorize family relations in different ways, and translations are often only approximate. There may be a single term for a certain complex relation in one language, but none in another. Even within a single language, there are sometimes ambiguities. In English, for example, let P be the relation of 'parent of', and S the relation of 'sister of'. Then $P \circ S$ is the relation of 'being a sister of a parent of'. This is certainly a *subrelation* of the relation of being an aunt of, but are the two relations identical? That depends on whether you include aunts by marriage, i.e. whether you regard the wives of the brothers of your parents as your aunts. If you *do* include them under the term, then $P \circ S$ is a proper subrelation of the relation of aunt. If you *don't* include them, then it is the whole relation, i.e. $P \circ S$ = 'aunt'.

2.3.4 Image

The last operation that we consider in this section records the action of a relation on a set. Let R be any relation from set A to set B, and let $a \in A$. We define the *image of a under R*, written $R(a)$, to be the set of all $b \in B$ such that $(a,b) \in R$.

EXERCISE 2.3.5 (WITH PARTIAL SOLUTION)

(a) Let $A = \{$John, Mary, Peter$\}$ and let $B = \{1,2,3\}$. Let R be the relation $\{($John,1$), ($Mary,2$), ($Mary,3$)\}$. What are the images of John, Mary, and Peter under R?

(b) Represent these three images in a natural way in the digraph for R.

(c) What is the image of 9 under the relation $\leq$ over the natural numbers?

Solution to (a) and (c):

(a) $R($John$) = \{1\}$, $R($Mary$) = \{2,3\}$, $R($Peter$) = \varnothing$.

(c) $\leq(9) = \{n \in \mathbf{N} : 9 \leq n\}$.

It is useful to 'lift' this notion from elements of A to subsets of A. If $X \subseteq A$ then we define the *image of X under R*, written $R(X)$, to be the set of all $b \in B$ such that $(x,b) \in R$ for some $x \in X$. In shorthand notation, $R(X) = \{b \in B : \exists x \in X, (x,b) \in R\}$.

EXERCISE 2.3.6

(a) Let A, B, R be as in the preceding exercise. Identify $R(X)$ for each one of the eight subsets X of A.

(b) What are the images of the following sets under the relation $\leq$ over the natural numbers? (i) $\{3,12\}$, (ii) $\{0\}$, (iii) $\varnothing$, (iv) the set of all evens, (v) the set of all odds, (vi) $\mathbf{N}$.

(c) Let P (for predecessor) be the relation defined by putting $(a,x) \in P$ iff $x = a+1$. What are the images of each of the above six sets under P?

(d) Identify the converse of the relations $\leq$ and P over the natural numbers, and the images of the above six sets under these converse relations.

It can happen that the notation $R(X)$ for the image of $X \subseteq A$ under R is ambiguous. This will be the case when the set A already has among its elements certain subsets X of itself. For this reason, some authors write $R``(X)$ instead. However, we will rarely be dealing with such sets, and will stick with the simpler notation.

2.4 Reflexivity and Transitivity

In this section and the following two, we will look at special properties that a relation may have (or fail to have). Some of these properties are useful when we ask questions about how *similar* two items are. Some are needed when we look for some kind of *order* among items. Two properties, reflexivity and transitivity, are essential to both tasks, and so we begin with them.

2.4.1 Reflexivity

Let R be a relation over a set A. We say that R is *reflexive* (over A) iff $(a,a) \in R$ for all $a \in A$. For example, the relation $\leq$ is reflexive over the natural numbers, since always $n \leq n$, but $<$ is not. Indeed, $<$ has the opposite property of being *irreflexive* over the natural numbers: never $n < n$.

Clearly, reflexivity and irreflexivity are not the only possibilities. A relation R over a set may be neither one nor the other – for example n is a prime divisor of n for some natural numbers (e.g. 2 is a prime divisor of itself) but fails for some others (e.g. 4 is not a prime divisor of 4).

If we draw a digraph for a reflexive relation, every point will have an arrow going from it to itself; an irreflexive relation will have no such arrows; a relation that is neither reflexive nor irreflexive will have such arrows for some but not all of its points. However, when a relation is reflexive we sometimes reduce clutter by omitting these arrows and treating them as understood.

EXERCISE 2.4.1 (WITH PARTIAL SOLUTION)

(a) Give another example of a relation over the natural numbers for each of the three properties reflexive/irreflexive/neither.

(b) Can a relation R ever be both reflexive and irreflexive over a set A? If so, when?

(c) Identify the status of the following relations as reflexive/irreflexive/ neither over the set of all people living in the UK: sibling of, shares at least one parent with, ancestor of, lives in the same city as, has listened to music played by.

(d) What does reflexivity mean in terms of a tabular representation of relations?

Solution to (c): The relation 'sibling of' as ordinarily understood is not reflexive, since we do not regard a person as a sibling of himself or herself. In fact, it is irreflexive. By contrast, 'shares at least one parent with' is reflexive. Its follows that these two relations are not quite the same. 'Ancestor of' is irreflexive'. 'Lives in the same city as', under it most natural meaning, is neither reflexive nor irreflexive – not everybody lives in a city. 'Has listened to music played by' is also neither reflexive nor irreflexive.

A subtle point needs attention. Let R be a relation over a set A, i.e. $R \subseteq A^2$. Then as we observed earlier, R is also a relation over any superset B of A, since then $R \subseteq A^2 \subseteq B^2$. It can happen that R is reflexive over A, but not over its superset B. This is because we may have $(a,a) \in R$ for all $a \in A$ but $(b,b) \notin R$ for some $b \in B-A$.

For this reason, whenever we say that a relation is reflexive or irreflexive, we should in principle specify what set A we have in mind. Doing this explicitly can be rather laborious, and so in practice the identification of A is often left as understood.

EXERCISE 2.4.2

(a) Give a small finite example of this dependence phenomenon.

(b) Show that if R is reflexive over A, and $B \subseteq A$, then the restriction of R to B is reflexive over B.

(c) Show that the intersection and union of any two reflexive relations are both reflexive, as is also the converse of any reflexive relation.

2.4.2 Transitivity

Another important property for relations is transitivity. We say that R is *transitive* iff whenever $(a,b) \in R$ and $(b,c) \in R$ then $(a,c) \in R$. For example, the relation $\leq$ over the natural numbers is transitive: whenever $a \leq b$ and $b \leq c$ then $a \leq c$. Likewise for the relation $<$.

But the relation of having some common prime factor is not transitive: 4 and 6 have a common prime factor (namely 2), and 6 and 9 have a common prime factor (namely 3), but 4 and 9 do not have any common prime factor. By the same token, the relation between people of being first cousins (i.e. sharing some grandparent, but not sharing any parent) is not transitive.

Another relation over the natural numbers failing transitivity is that of being an immediate predecessor: 1 is an immediate predecessor of 2, which is an immediate predecessor of 3, but 1 does not stand in that relation to 3. In this example, the relation is indeed *intransitive*, in the sense that whenever $(a,b) \in R$ and $(b,c) \in R$ then $(a,c) \notin R$. In contrast, the relation of sharing a common prime factor, and that of being a first cousin, are clearly neither transitive nor intransitive.

In an unabbreviated digraph for a transitive relation, whenever there is an arrow from one point to a second, and another arrow from the second point to a third, there should also be an arrow from the first to the third. Evidently, this tends to clutter the picture. So often, when a relation is transitive we adopt the convention of omitting the 'third arrows', treating them as understood. In any particular case, one must be clear about the convention one is following.

EXERCISE 2.4.3

(a) Give another example of a relation over the natural numbers for each of the three properties transitive/intransitive/neither.

(b) Can a relation R ever be both transitive and intransitive? If so, when?

(c) Draw a table indicating the status of each of the following relations on the transitive/intransitive/neither dimension: sister of, sibling of, parent of, ancestor of.

(d) Show that the intersection of any two transitive relation is transitive.

(e) Give an example to show that the union of two transitive relations need not be transitive.

2.5 Equivalence Relations and Partitions

We now focus on properties that are of particular interest when we want to express a notion of similarity or equivalence.

2.5.1 Symmetry

We say that R is *symmetric* iff whenever $(a,b) \in R$ then $(b,a) \in R$. For example, the relation of identity (alias equality) over the natural numbers is symmetric: whenever $a = b$ then $b = a$. So is the relation of sharing a common prime factor: if a has some prime factor in common with b, then b has a prime factor (indeed, the same one) in common with a. On the other hand, neither $\leq$ nor $<$ over the natural numbers is symmetric.

The way in which $<$ fails symmetry is not the same as the way in which $\leq$ fails it. When $n < m$ then we *never* have $m < n$, and the relation is said to *asymmetric*. In contrast, when $n \leq m$ we sometimes have $m \leq n$ (when $m = n$) but sometimes $m \not\leq n$ (when $m \neq n$), and so this relation is neither symmetric nor asymmetric.

In an unabbreviated digraph for a symmetric relation, whenever there is an arrow from one point to a second, then there is an arrow going back again. Here too, a special convention can be introduced to reduce clutter: put a head at each end of the arrow or, better, replace the arrows by links without heads.

EXERCISE 2.5.1

(a) What does symmetry mean in terms of the tabular representation of a relation?

(b) Give another example of a relation over the natural numbers for each of the three properties symmetric/asymmetric/neither.

(c) Can a relation R ever be both symmetric and asymmetric? If so, when?

(d) Determine the status of each of the following relations on the transitive/intransitive/neither dimension: brother of, sibling of, parent of, ancestor of.

(e) Show that the converse of any symmetric relation is symmetric, as are also the intersection and union of any symmetric relations.

2.5.2 Equivalence Relations

When a relation is both reflexive and symmetric, it is sometimes called a *similarity relation*. When it has all three properties – transitivity, symmetry, and reflexivity – it is called an *equivalence relation*.

Equivalence relations are often written as using a symbol such as $\approx$ to bring out the idea that they behave rather like identity. And like identity they are usually written by *infixing*, that is as $a \approx b$, rather than in *basic set notation* as in $(a,b) \in R$ or by *prefixing* as in Rab.

EXERCISE 2.5.2 (WITH PARTIAL SOLUTION)

(a) Give another example of an equivalence relation over the natural numbers, and one over the set of all polygons in geometry.

(b) (i) Check that the identity relation over a set is an equivalence relation over that set. (ii) Show also that it is the *least* equivalence over that set, in the sense that it is included in every equivalence relation over the set. (iii) What is the largest equivalence relation over a set?

(c) Over a set of people, is the relation of having the same nationality an equivalence relation?

(d) What does the digraph of an equivalence relation look like?

Solutions to (a–c):

(a) For example, the relation of having the same parity (i.e. both even, or else both odd) is an equivalence relation. For polygons, the relation of having the same number of sides is an equivalence relation. Of course, there are many others in both domains.

(b) (i) Identity is reflexive over any set because always $a = a$. It is symmetric because whenever $a = b$ then $b = a$. It is transitive because whenever $a = b$ and $b = c$ then $a = c$. (ii) It is included in every other equivalence relation $\approx$ over the same set, because we have $a \approx a$ for every $a \in A$, so $a \approx b$ whenever $a = b$. (iii) The largest equivalence relation over A is A^2. Indeed it is the largest relation over A^2, and it is easily checked to be an equivalence relation.

(c) In the case that everyone in A has exactly one nationality, then this is an equivalence relation. But if someone in A has no nationality, then reflexivity fails, and if someone has more than one nationality, transitivity may fail. In contrast, the relation of 'having the same *set of nationalities*' is always an equivalence relation.

Alice Box: Identity, equality, replacement

Alice: You speak of people being identical to each other, but for numbers and other mathematical objects I more often hear of equality. Are these the same?

Hatter: Logicians tend to speak of identity while number theorists talk of equality, but they are the same. Identity is equal to equality, equality is identical with identity. However, both terms are also sometimes used, rather loosely, to indicate any reasonably 'tight' equivalence relation.

Alice: Is there any obvious difference between identity and other equivalence relations, apart from being the smallest?

Hatter: Yes, there is a very important difference, which is a consequence of being the smallest. When $a = b$, they are elements of exactly the same sets and

(*Continued*)

Alice Box: (Continued)

have exactly the same properties. Any mathematical statement that is true of one is true of the other, so *we may replace one by the other in any statement without loss of truth.* This is, along with reflexivity, is one of the fundamental properties of identity in first-order logic.

Alice: Can you give me an example?

Hatter: We did just that in one of the parts of the last exercise, when we said that whenever $a \approx a$ then if $a = b$ we have $a \approx b$. Here we were replacing the second a in $a \approx a$ by b. In school you learned this as 'substitution of equals gives equals'.

2.5.3 Partitions

Your hardcopy correspondence is in a mess – one big heap. You need to classify it, putting items into mutually exclusive but together exhaustive categories, but without going into subcategories. Mathematically speaking, this means that you want to create a *partition* of the set of all the items in the heap.

   We now define this concept. Let A be any non-empty set, and let $\{B_i\}_{i \in I}$ be a collection of subsets of A.

- We say that the collection *exhausts* A iff $\cup \{B_i\}_{i \in I} = A$, i.e. iff every element of A is in at least one of the B_i. Using the notation that we introduced earlier for the universal and existential quantifiers, iff $\forall a \in A \; \exists i \in I \, (a \in B_i)$.

- We say that the collection is *pairwise disjoint* iff every two distinct sets B_i in the collection are disjoint. That is, for all $i, j \in I$, if $B_i \neq B_j$ then $B_i \cap B_j = \varnothing$. Using logical notation, iff $\forall i, j \in I \, (B_i \neq B_j \rightarrow B_i \cap B_j = \varnothing)$.

   A *partition of A* is defined to be any collection $\{B_i\}_{i \in I}$ of non-empty subsets of A that are pairwise disjoint and together exhaust A. The sets B_i are called the *cells* (or sometimes, *blocks*) of the partition. We can diagram a partition in the following manner.

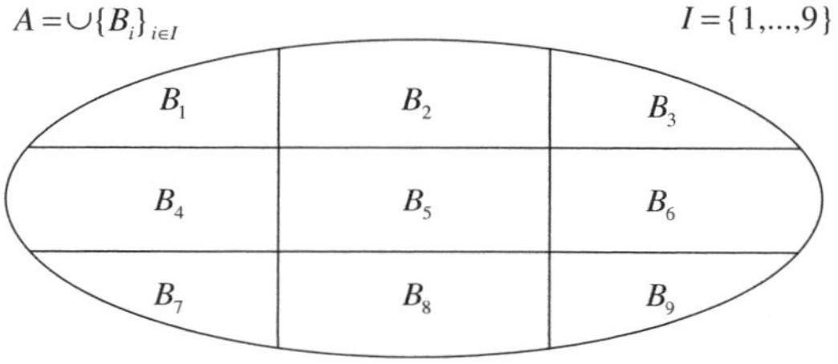

Figure 2.2   *Diagram for a partition.*

EXERCISE 2.5.3 (WITH SOLUTION)

(a) Which of the following are partitions of $A = \{1,2,3,4\}$? (i) $\{\{1,2\}, \{3\}\}$, (ii) $\{\{1,2\}, \{2,3\}, \{4\}\}$, (iii) $\{\{1,2\}, \{3,4\}, \varnothing\}$, (iv) $\{\{1,4\}, \{3,2\}\}$.

(b) We say that one partition of a set is *at least as fine* as another, iff every cell of the former is a subset of a cell of the latter. What is the finest partition of $A = \{1,2,3,4\}$? What is the least fine partition? In general, if A has $n \geq 1$ elements, how many cells in its finest and least fine partitions?

Solution:

(a) Only (iv) is a partition of A. In (i) the cells do not exhaust A, in (ii) they are not pairwise disjoint, and in (iii) one cell is the empty set, which is not allowed.

(b) The finest partition of $A = \{1,2,3,4\}$ is $\{\{1\}, \{2\}, \{3\}, \{4\}\}$, and the least fine is $\{\{1, 2, 3, 4\}\}$. In general, if A has $n \geq 1$ elements, then its finest partition has n cells, while its least fine partition has only 1 cell.

2.5.4 The Correspondence Between Partitions and Equivalence Relations

It turns out that partitions and equivalence relations are two sides of the same coin. On the one hand, every partition of a set A determines an equivalence relation over A in a natural manner; on the other hand, every equivalence relation over a non-empty set A determines, in an equally natural way, a partition over A.

The verification of this is rather abstract, and hence challenging, but you should be able to follow it. We begin with the left-to-right direction. Let A be any set, and let $\{B_i\}_{i \in I}$ be a partition of A. We define the relation R *associated with* this partition by putting $(a,b) \in R$ iff a and b are in the same cell, i.e. iff $\exists i \in I$ with $a,b \in B_i$. We need to show that it is an equivalence relation.

- R is clearly *reflexive* over A: since the partition exhausts A, $\forall a \in A$, $\exists i \in I$ with $a \in B_i$ and so immediately $\exists i \in I$ with both of $a,a \in B_i$.

- Equally clearly, R is *symmetric*: when $(a,b) \in R$ then $\exists i \in I$ with $a,b \in B_i$ so immediately $b,a \in B_i$ and thus $(b,a) \in R$.

- Finally, R is *transitive*. Suppose $(a,b) \in R$ and $(b,c) \in R$; we want to show $(a,c) \in R$. Since $(a,b) \in R$, $\exists i \in I$ with both $a,b \in B_i$, and since $(b,c) \in R$, $\exists j \in I$ with $b,c \in B_j$. But since the cells of the partition are pairwise disjoint, either

$B_i = B_j$ or $B_i \cap B_j = \varnothing$. Since $b \in B_i \cap B_j$ the latter is not an option, so $B_i = B_j$. Hence both $a,c \in B_i$, which gives us $(a,c) \in R$ as desired.

For the right-to-left direction, Let $\approx$ be an equivalence relation over a non-empty set A. For each $a \in A$ we consider its image $\approx(a) = \{x \in A : a \approx x\}$ under the relation $\approx$. This set is usually written as $|a|_\approx$ or simply as $|a|$ when the equivalence relation $\approx$ is understood, and that is the notation we will use here.

- *Non-emptiness.* Since $\approx$ is a relation over A, each set $|a|$ is a subset of A, and it is non-empty because $\approx$ is reflexive over A.

- *Exhaustion.* The sets $|a|$ together exhaust A, i.e. $\cup\{|a|\}_{a \in A} = A$, again because $\approx$ is reflexive over A.

- *Pairwise disjointedness.* Let $a,a' \in A$ and suppose $|a| \cap |a'| \neq \varnothing$. We show that $|a| = |a'|$ (contrapositive argument again). For that, it suffices to show that $|a| \subseteq |a'|$ and conversely $|a'| \subseteq |a|$. We do the former; the latter is similar. Let $x \in |a|$ i.e. $a \approx x$; we need to show that $x \in |a'|$, i.e. that $a' \approx x$. By the initial supposition, there is a y with $y \in |a| \cap |a'|$, so $y \in |a|$ and $y \in |a'|$, so $a \approx y$ and $a' \approx y$. Since $a \approx y$ symmetry gives us $y \approx a$, and so we may apply transitivity twice: first to $a' \approx y$ and $y \approx a$ to get $a' \approx a$, and then to that and $a \approx x$ to get $a' \approx x$ as desired.

Note that in the proof of the right-to-left part, we appealed to all three of the conditions reflexivity, symmetry and transitivity. You can't get away with less.

When $\approx$ is an equivalence relation over A, the sets $|a|$ for $a \in A$ (i.e. the cells of the corresponding partition) are called *equivalence classes* of $\approx$, and are thus the same as the *cells of the corresponding partition*.

EXERCISE 2.5.4

(a) Let A be the set of all positive integers from 1 to 10. Consider the partition into evens and odds. Write this partition by enumeration as a collection of sets, then describe the corresponding equivalence relation in words, and write it by enumeration as a set of ordered pairs.

(b) Let A be the set of all positive integers from 2 to 16. Consider the relation of having exactly the same prime factors (so e.g. $6 \approx 12$ since they have the same prime factors 2 and 3). Identify the associated partition by enumerating it as a collection of subsets of A.

(c) How would you describe the equivalence relation associated with the finest (respectively: least fine) partition of A?

2.6 Relations for Ordering

Suppose that we want to use a relation to order things. It is natural to require it to be transitive. We can choose it to be reflexive, in which case we get an *reflexive order* (like $\leq$ over the natural numbers or $\subseteq$ between sets); or to be irreflexive, in which case we get what is usually known as a *strict order* (such as $<$ or $\subset$). In this section we examine some special kinds of inclusive order, and then look at strict orders.

2.6.1 Partial Order

Consider a relation that is both reflexive and transitive. What other properties do we want it to have if it is to serve as an ordering?

Not symmetry, for that would make it an equivalence relation. Asymmetry? Not quite, for a reflexive relation over a non-empty set can never be asymmetric. Recall the reason: if A is non-empty, then there is some $a \in A$, so reflexivity gives $(a,a) \in R$, so to repeat ourselves both $(a,a) \in R$ and $(a,a) \in R$, contrary to asymmetry.

What we need is the closest possible thing to asymmetry: whenever $(a,b) \in R$, then $(b,a) \notin R$ provided that $a \neq b$. This property is known as *antisymmetry*, and it is usually formulated in a contraposed (and thus equivalent) manner: whenever $(a,b) \in R$ and $(b,a) \in R$ then $a = b$.

A relation R over a set A that is reflexive (over A), transitive, and also antisymmetric is called a *partial order* (or partial ordering) of A, and the pair (A,R) is called for short a *poset*. It is customary to write a partial ordering as $\leq$ (or some square or curly variant of the same sign) even though, as we will soon see, the familiar relation 'less than or equal to' over the natural numbers (or any other number system) has a further property that not all partial orderings share.

Two examples of partial order:

- For sets, the relation $\subseteq$ of inclusion is a partial order. As we already know, it is reflexive and transitive. We also have antisymmetry, since whenever $A \subseteq B$ and $B \subseteq A$ then $A = B$.

- In arithmetic, an important example is the relation of being a divisor of, over the positive integers, i.e. the relation R over $\mathbf{N}^+$ defined by $(a,b) \in R$ iff $b = ka$ for some $k \in \mathbf{N}^+$.

EXERCISE 2.6.1

    (a) Check that the relation of being divisor of, over the positive integers, is indeed a partial order.

(b) What about the corresponding relation over $\mathbf{Z}$, i.e. the relation R over $\mathbf{Z}$ defined by $(a,b) \in R$ iff $b = ka$ for some $k \in \mathbf{Z}$?

(c) And the corresponding relation over $\mathbf{Q}^+$, i.e. the relation R over $\mathbf{Q}^+$ defined by $(a,b) \in R$ iff $b = ka$ for some $k \in \mathbf{Q}^+$?

(d) Consider the relation R over people defined by $(a,b) \in R$ iff either $b = a$ or b is descended from a. Is it a poset?

(e) Show that the relation of 'being at least as fine as', between partitions of a given set A, is a partial order.

2.6.2 Linear Orderings

A relation R over a set A is said to be *complete* over A iff for all $a,b \in A$, either $(a,b) \in R$ or $(b,a) \in R$. A poset that is also complete is often called a *linear* (or *total*) *ordering*.

Clearly the relation $\leq$ over $\mathbf{N}$ is complete and so a linear ordering, since for all $m,n \in \mathbf{N}$ either $m \leq n$ or $n \leq m$. So is the usual lexicographic ordering of words in a dictionary.

On the other hand, whenever a set A has more than one element, then the relation $\subseteq$ over $\mathcal{P}(A)$ is not complete. Reason: take any two distinct $a,b \in A$, and consider the singletons $\{a\}$, $\{b\}$; they are both elements of $\mathcal{P}(A)$, but neither $\{a\} \subseteq \{b\}$ nor $\{b\} \subseteq \{a\}$ because $a \neq b$.

EXERCISE 2.6.2 (WITH SOLUTION)

(a) Give two more linear orderings of $\mathbf{N}$.

(b) Which of the following are linear orderings over an arbitrary set of people? (i) is at least as old as, is (ii) identical to or a descendent of.

Solution:

(a) There are plenty, but here are two: (i) the relation $\geq$, i.e. the converse of $\leq$, (ii) the relation that puts all odd positive integers first, and then all the even ones, each of these blocks ordered separately as usual.

(b) (i) Yes, it meets all the requirements, (ii) no, since it is not complete.

2.6.3 Strict Orderings

Whenever we have a reflexive transitive relation $\leq$ we can always look at what is known as its *strict part*. It is usually written as $<$ (or a square or curly variant of this) even though it need not have all the properties of 'less than' over the usual number systems. The definition is as follows: $a < b$ iff $a \leq b$ but $a \neq b$. In language that looks like gibberish but makes sense if you read it properly: $(<) = (\leq \cap \neq)$.

EXERCISE 2.6.3 (WITH SOLUTION)

    (a) Show that every asymmetric relation over a set A is irreflexive.

    (b) Show that when $\leq$ is a partial ordering over a set A then its strict part $<$ is asymmetric and transitive.

Solution:

    (a) Suppose that $<$ is asymmetric, but not irreflexive. We get a contradiction. Since the relation is not irreflexive, there is an $a \in A$ with $a < a$. Hence by asymmetry, not $a < a$, giving us a contradiction and we are done.

    (b) For asymmetry, suppose $a < b$ and $b < a$; we get a contradiction. By the suppositions, $a \leq b$, $b \leq a$ and $a \neq b$, which is impossible by the antisymmetry of $\leq$.

For transitivity, suppose $a < b$ and $b < c$; we want to show that $a < c$. By the suppositions, $a \leq b$ and $b \leq c$ but $a \neq b$ and $b \neq c$. Transitivity of $\leq$ thus gives $a \leq c$; it remains to check that $a \neq c$. Suppose $a = c$; we get a contradiction. Since $b \leq c$ and $a = c$ we have $b \leq a$, so by the antisymmetry of $\leq$ using also $a \leq b$ we have $a = b$, giving us the desired contradiction with $a \neq b$.

Alice Box: Proof by contradiction (reductio ad absurdum)

Alice: There is something in the solution to this exercise that worries me. We supposed the *opposite* of what we wanted to show. For example, to show that the relation $<$ there is asymmetric, we supposed that both $a < b$ and $b < a$, which is the negation of what we are trying to prove.

Hatter: Indeed we did, and the goal of the argument changed when we made the supposition: it became one of deriving a contradiction. In the exercise, we got our contradictions very quickly; sometimes it takes more argument.

(Continued)

Alice Box: (Continued)

Alice: Will any contradiction do?

Hatter: Any contradiction you like. Any pair of propositions α and $\neg\alpha$. Such pairs are as bad as each other, and thus just as good for the purposes of the proof.

Alice: Is this procedure some new invention of modern logicians?

Hatter: Not at all. It was well known to the ancient Greeks, and can be found in Euclid. In the Middle Ages it was taught under its Latin name *reductio ad absurdum*, sometimes abbreviated to *RAA*, and this name is still often used.

Alice: Can we always apply it?

Hatter: Any time you like. Some people like to use whenever they can, others do so only when they get stuck without it. Most are somewhere in the middle – they use it when they see that it can make the argument visibly shorter or more transparent.

Alice: Does it do that for the last exercise?

Hatter: That's for you to judge. Do the same exercise again without using proof by contradiction, and compare the answers.

In the preceding chapter we observed that the relation of immediate inclusion between finite sets may be represented by a Hasse diagram. The same is true for any partial order $\leq$ over a finite set. The links of the diagram, read from below to above, represent the relation of *being an immediate predecessor of*. This is the relation that holds between a and b iff $a < b$ but there is no x with $a < x < b$, where $<$ is the strict part of $\leq$.

Once we have a Hasse diagram for the immediate predecessor part of a partial ordering, we can read off from it the entire relation: $a \leq b$ iff there is an ascending path (with at least one element) from a to b. Evidently, this is a much more economical representation than drawing the digraph for the entire partial ordering. For this reason, we loosely call it the Hasse diagram of the partial ordering itself, and similarly for its strict part.

EXERCISE 2.6.4

Draw a Hasse diagram for (the immediate predecessor part of) the relation of being an exact divisor of, over the set of positive integers up to 13.

We have seen in an exercise that when $\leq$ is a partial ordering over a set A then its strict part $<$ is asymmetric and transitive. We also have a converse:

whenever a relation $<$ is both asymmetric and transitive, then the relation $\leq$ defined by putting $a \leq b$ iff either $a < b$ or $a = b$ is a partial order. Given this two-way connection, asymmetric transitive relations are often called *strict partial orders*.

EXERCISE 2.6.5

> Show, as claimed in the text, that when $<$ is a transitive asymmetric relation, then the relation $\leq$ defined by putting $a \leq b$ iff either $a < b$ or $a = b$ is a partial order.

2.7 Closing with Relations

2.7.1 Transitive Closure of a Relation

Suppose you are given a relation R that is not transitive, but which you want to 'make' transitive. Of course, you cannot change the status of R itself, but you can expand it to a larger relation that satisfies transitivity.

In general, there will be many of these. For example whenever $R \subseteq A^2$ then A^2 itself is a transitive relation that includes R. But there will be much smaller ones, and it is not difficult to see that there must always be a *unique smallest* transitive relation that includes R; it is called the *transitive closure* of R and is written as R^*. It may be defined as the intersection of the collection of all transitive relations that include R. That is: when R is a relation over A then $R^* = \cap \{S : R \subseteq S \subseteq A^2$ and S is transitive$\}$.

EXERCISE 2.7.1 (WITH PARTIAL SOLUTION)

> (a) Identify the transitive closures of the following relations: (i) parent of, (ii) mother of, (iii) descendant of, (iv) sister of.
>
> (b) Show that R^*, as defined, is claimed the least transitive relation that includes R, i.e. that (i) $R \subseteq R^*$, (ii) R^* is transitive, and (iii) $R^* \subseteq S$ for every transitive relation S with $R \subseteq S$.
>
> (c) Write the following assertions in the notation for converse and transitive closure, and determine whether they are true or false: (i) the converse of the transitive closure of a relation equals the transitive closure of the converse of that relation, (ii) the intersection of the

transitive closures of two relations equals the transitive closure of their intersection.

Solution to (a): Ancestor of. (ii) Ancestor of in the female line (there is no single word for this in English). (iii) Descendant of (as this relation is already transitive, it coincides with its transitive closure). (iv) This one is tricky. Remember that a female will be a sister of each of her sisters, so that for her the transitive closure will include selfhood. One way of expressing this relation is as follows, writing $(a,b) \in S$ for a is a sister of b, and F for the set of female persons: $S^* = S \cup \{(y,y) : y \in F$ and $\exists z$ $(z,y) \in S\}$.

As well as this 'top-down' definition using intersection, one can work with a 'bottom-up' one using union. We define the relations R_0, $R_1, \ldots$ as follows:

$$R_0 = R$$
$$R_{n+1} = R_n \cup \{(a,c) : \exists x \text{ with } (a,x) \in R_n \text{ and } (x,c) \in R\}$$
$$R^* = \cup \{R_n : n \in \mathbf{N}\}.$$

In this way, R^* is built up by successively adding pairs (a,c) whenever (a,x) is in the relation constructed so far and $(x,c) \in R$. We keep on doing this until there are no such pairs (a,c) still needing to be put in, and keep going indefinitely if there are always such pairs.

2.7.2 Closure of a Set Under a Relation

Transitive closure is in fact a particular instance of a more general construction of great importance – the closure of an arbitrary set under an arbitrary relation.

To introduce it, we recall the definition of image from earlier in this chapter. The *image* $R(X)$ of a set X under a relation R is the set of all b such that $(x,b) \in R$ for some $x \in X$; briefly $R(X) = \{b : \exists x \in X, (x,b) \in R\}$.

Now, suppose we are given a relation R (not necessarily transitive) over a set B, and a subset $A \subseteq B$. We define the *closure $R[A]$ of A under R* to be the least subset of B that includes A and also includes $R(X)$ whenever it includes X.

Again, this is a top-down definition, with 'least' understood as the result of intersection. It can also be expressed bottom-up, as follows:

$$A_0 = A$$
$$A_{n+1} = A_n \cup R(A_n) \text{ for each natural number } n$$
$$R[A] = \cup \{A_n : n \in \mathbf{N}\}.$$

The closure of A under R is thus constructed by beginning with A itself, and at each stage adding in the elements of the image of the set so far constructed.

EXERCISE 2.7.2 (WITH SOLUTION)

In the set $\mathbf{N}$ of natural numbers, what is the closure $R[A]$ of the set $A = \{2,5\}$ under the following relations: (a) $(m,n) \in R$ iff $m,n \in \mathbf{N}$ and $n = m+1$, (b) $(m,n) \in R$ iff $m,n \in \mathbf{N}$ and $n = m{-}1$, (c) $(m,n) \in R$ iff $m,n \in \mathbf{N}$ and $n = 2m$, (d) $(m,n) \in R$ iff $m,n \in \mathbf{N}$ and $n = m/2$, (e) $\leq$, (f) $<$.

Solution: (a) $\{n \in \mathbf{N} : n \geq 2\}$, (b) $\{n \in \mathbf{N} : n \leq 5\} = \{0,1,2,3,4,5\}$, (c) $\{2,4,8,\ldots ; 5,10,20,\ldots\}$, (d) $\{1,2,5\}$, (e) $\{n \in \mathbf{N} : n \geq 2\}$, (f) $\{n \in \mathbf{N} : n \geq 2\}$.

Comment: If you wrote $\{n \in \mathbf{N} : n > 2\}$ as your answer to (f), you forgot that $A \subseteq R[A]$ for every relation R.

Notation and terminology: When the relation R is understood, we sometimes write the closure $R[A]$ more briefly as A^{+}, and say that R *generates* the closure from A.

FURTHER EXERCISES

2.1. *Cartesian products*

(a) Show that $A{\times}(B{\cap}C) = (A{\times}B){\cap}(A{\times}C)$.

(b) Show that $A{\times}(B{\cup}C) = (A{\times}B){\cap}(A{\times}C)$.

2.2. *Domain, range, join, composition, image*

(a) Consider the relations $R = \{(1,7), (3,3), (13,11)\}$ and $S = \{(1,1), (1,7), (3,11), (13,12), (15,1)\}$ over the positive integers. Identify $dom(R{\cap}S)$, $range(R{\cap}S)$, $dom(R{\cup}S)$, $range(R{\cup}S)$.

(b) In the same example, identify $join(R,S)$, $join(S,R)$, $S{\circ}R$, $R{\circ}S$, $R{\circ}R$, $S{\circ}S$.

(c) In the same example, identify $R(X)$ and $S(X)$ for $X = \{1,3,11\}$ and $X = \varnothing$.

2.3. *Reflexivity and transitivity*

(a) Show that R is reflexive over A iff $I_A \subseteq R$. Here I_A is the identity relation over A, defined in an earlier exercise.

(b) Show that the converse of a reflexive relation R over a set A is reflexive over A.

(c) Show that R is transitive iff $R \, R \subseteq R$.

(d) In a previous exercise we showed that the intersection of any two transitive relations is transitive but their union may not be. (i) Show that the converse and composition of transitive relations are also transitive. (ii) Give an example to show that the complement of a transitive relation need not be transitive.

(e) A relation R is said the be *acyclic* iff there are no $a_1, \ldots, a_n$ ($n \geq 2$) such that each $(a_i, a_{i+1}) \in R$ and also $a_n = a_1$. (i) Show that a transitive irreflexive relation is always acyclic. (ii) Show that every acyclic relation is irreflexive. (iii) Give an example of an acyclic relation that is not transitive.

2.4. *Symmetry, equivalence relations and partitions*

(a) Show that the following three conditions are equivalent: (i) R is symmetric, (ii) $R \subseteq R^{-1}$, $R = R^{-1}$.

(b) Show that a reflexive relation over a non-empty set can never be asymmetric.

(c) Show that if R is reflexive over A and also transitive, then the relation S defined by $(a,b) \in S$ iff both $(a,b) \in R$ and $(b,a) \in R$ is an equivalence relation.

(d) Show that the intersection of two equivalence relations is an equivalence relation, but that this is not the case for unions. *Hint*: make use of the results of exercises on reflexivity, transitivity, and symmetry in this chapter.

(e) Enumerate all the partitions of $A = \{1,2,3\}$ and draw a Hasse diagram for them under fineness.

(f) Show that one partition of a set is at least as fine as another iff the equivalence relation associated with the former is a subrelation of the equivalence relation associated with the latter.

2.5. *Antisymmetry, partial order, linear order*

(a) Let R be any reflexive, transitive relation over a set A. Define S over A by putting $(a,b) \in S$ iff either $a = b$ or both $(a,b) \in R$ and $(b,a) \notin R$. Show that S partially orders A.

(b) Show that the converse of a linear order is linear, but that the intersection and composition of two linear orders need not be linear.

(c) Show that the identity relation over a set A is the unique partial order of A that is also an equivalence relation.

2.6. *Strict orderings*

(a) Give examples of relations that are (i) transitive but not asymmetric, and (ii) asymmetric but not transitive.

(b) Show that a relation is antisymmetric iff its strict part is asymmetric.

(c) Let A be a set and $\leq$ a partial ordering of A. An element $a \in A$ is said to be a *minimal* element of A (under $\leq$) iff there is no $b \in A$ with $b < a$. On the other hand, an element $a \in A$ is said to be a *least* element of A (under $\leq$) iff $a \leq b$ for every $b \in A$. Show the following for sets A and partial orderings $\leq$: (i) whenever a is a least element of A then it is a minimal element of A, (ii) The converse can fail (give a simple example); (iii) A can have zero, one, or more than one minimal elements (give an example of each); (iv) A can have at most one least element under a partial ordering, i.e. if a least element exists then it is unique.

2.7. *Closure*

(a) Show that always $X \cup R(X) \subseteq R[X]$.

(b) Show that $R[A] = R(A)$ if R is both reflexive over A and transitive.

Selected Reading

As for the preceding chapter, a classic of beautiful exposition, but short on exercises:

Paul R. Halmos *Naive Set Theory*. Springer, 2001 (new edition), Chapters 6–7.

The material is covered with lots of exercises in:

Seymour Lipschutz *Set Theory and Related Topics*. McGraw Hill Schaum's Outline Series, 1998, Chapter 3.

All textbooks on discrete mathematics have something on relations, although it is sometimes spread out in different chapters. One popular text is:

Richard Johnsonbaugh *Discrete Mathematics*. Pearson, 2005 (sixth edition) Chapter 3.

3

Associating One Item with Another: Functions

Chapter Outline

Functions occur everywhere in mathematics and computer science. In this chapter we introduce the basic concepts needed in order to work with them.

We begin by explaining the intuitive idea of a *function* and its mathematical definition as a special kind of relation. We then we see how the concepts for relations that were studied in the previous chapter unfold in this case (*domain, range, image, restriction, closure, composition, inverse*), and distinguish some important kinds of function (*injective, surjective, bijective*) with special behaviour. These concepts permit us to link functions with counting, with the *equinumerosity, comparison* and surprisingly versatile *pigeonhole* principles. Finally we identify some very simple functions that appear over and again (*identity, constant, projection* and *characteristic* functions), as well as the deployment of functions to represent *sequences* and *families*.

3.1 What is a Function?

Traditionally, a function was seen as a rule, often written as an equation, which associates any number (called the *argument* of the function) with another number, called the *value* of the function. The concept has for long been used in

D. Makinson, *Sets, Logic and Maths for Computing*,
DOI: 10.1007/978-1-84628-845-6_3, © Springer-Verlag London Limited 2008

physics and other sciences to describe processes whereby one quantity (such as the temperature of a gas, the speed of a car) affects another (its volume, its braking distance). In accord with such applications, the argument of the function was sometimes called the 'independent variable', and the value of the function termed the 'dependent variable'. The idea is that each choice for the argument or independent variable causally determines a value for the dependent variable.

Over the last two hundred years, the concept of a function evolved in the direction of greater abstraction and generality. The argument and value of the function need not be numbers – they can be items of any kind whatsoever. The function need not reflect a causal relationship, nor indeed any physical process, although these remain important applications. The function may not even be expressible by any linguistic rule, although all of the functions that we will encounter in this book are. Taken to the limit of abstraction, a function is no more than a set of ordered pairs, i.e. a relation in the sense of the preceding chapter, which satisfies a certain additional condition.

The formal definition is as follows. A *function* from a set A to a set B is any binary relation R from A to B such that *for all $a \in A$ there is exactly one* $b \in B$ with $(a,b) \in R$. The italicised parts of the definition deserve special attention.

- '*Exactly one. . .*': This implies that there is always *at least one* $b \in B$ with $(a,b) \in R$, and *never more* than one.

- '*For all $a \in A$. . .*': This implies that the specified source A of the relation is in fact its *domain*: there is no element a of A that fails to be the first term in some pair $(a,b) \in R$.

Strictly speaking, this is the definition of a *one-place* function, from a *single set* A to a set B. However, a function can have more than one argument, and we can generalize the definition accordingly. If $A_1, . . ., A_n$ are sets, then an n-place function from $A_1, . . ., A_n$ is an $(n+1)$-place relation R such that for all $a_1, . . ., a_n$ with each $a_i \in A_i$ there is exactly one $b \in B$ with $(a_1, . . ., a_n, b) \in R$.

In appearance, this is more general than the definition of one-place functions, which comes down to the case where $n = 1$. But in fact, it is already covered by that case: we may treat an n-place function from $A_1, . . ., A_n$ as a one-place function from the Cartesian product $A_1 \times . . . \times A_n$ to B. For example, addition and multiplication on the natural numbers are usually thought of a two-place functions f and g with $f(x,y) = x+y$ and $g(x,y) = x \cdot y$, but they may also be thought of as one-place functions on $\mathbf{N} \times \mathbf{N}$ into $\mathbf{N}$ with $f((x,y)) = x+y$ and $g((x,y)) = x \cdot y$. So there will be no real loss in generality when, to keep notation simple, we formulate principles in terms of one-place functions.

From time to time we will make use of a generalization of the notion of a function that relaxes the definition: A *partial function* from a set A to a set B is a binary relation R from A to B such that for all $a \in A$ there is *at most one* $b \in B$ with $(a,b) \in R$. The difference is that there may be elements of A that do not have the relation R to any element of B. Partial functions are also important for computer science, but for the present we will focus on functions in the full sense of the term.

In terms of tables: a function from A to B is a relation whose table has exactly one 1 in each row. In terms of digraphs: every point in the A circle has exactly one arrow going out from it to the B circle.

EXERCISE 3.1.1 (WITH PARTIAL SOLUTION)

For each of the following relations from $A = \{a,b,c,d\}$ to $B = \{1,2,3,4,5\}$, determine whether or not it is a function from A to B in all three ways – via the definition, the table and the digraph. Whenever your answer is negative, give your reason. Are any of the relations that are not functions from A to B nevertheless partial functions from A to B?

(a) $\{(a,1), (b,2), (c,3)\}$

(b) $\{(a,1), (b,2), (c,3), (d,4), (d,5)\}$

(c) $\{(a,1), (b,2), (c,3), (d,5)\}$

(d) $\{(a,1), (b,2), (c,2), (d,1)\}$

(e) $\{(a,5), (b,5), (c,5), (d,5)\}$

Partial solution: We solve in terms of the definition. (a) No, since $d \in A$ but there is no pair of the form (d,x) in the relation, so that the 'at least one' condition fails. It is nevertheless a partial function from A to B. (b) No, since both $(d,4)$ and $(d,5)$ are in the relation and evidently $4 \neq 5$ (the 'at most one' condition fails). For this reason, it is not even a partial function from A to B. (c) Yes: each element of the source is related to exactly one element of the target is; it does not matter that the element 4 of the target 'left out'. (d) Yes; it does not matter that the element 2 of the target is 'hit twice'. (e) Yes; it does not matter that the only element of the target that is hit is 5. This is called the *constant function* with value 5 from A to B.

Functions are usually referred to with lower case letters $f,g,h,\ldots$ Since for all $a \in A$ there is a unique $b \in B$ with $(a,b) \in f$, we may use a very convenient notation for them. We call that unique b the *value* of a under the function f, and write it as $f(a)$. In computer science we also say that $f(a)$ is the *output* of the function f for input a.

For brevity, we also write $f: A \to B$ to mean that f is a function from A to B. In the case that $B = A$, so that we have a function $f: A \to A$ from A to A, we usually say briefly that f is a function *on* A.

3.2 Operations on Functions

As functions are relations, all the operations that we introduced in the theory of relations apply to them. However, some are more useful than others in this context, or may be expressed in a novel way, and we begin by reviewing them. At times, terminology is used a little differently from the way it is employed in the theory of relations; we will note these occasions as they arise.

3.2.1 Domain and Range

These two concepts carry over without change. Recall that when R is a relation from A to B, then $dom(R) = \{a \in A : \exists b \in B \,((a,b) \in R)\}$. When R is in fact a function f, this reduces to $dom(f) = \{a \in A : \exists b \in B \,(f(a) = b)\}$. Thus when $f: A \to B$, then $dom(f) = A$.

Likewise, $range(R) = \{b \in B : \exists a \in A \,((a,b) \in R)\}$, which for functions reduces to $range(f) = \{b \in B : \exists a \in A \,(f(a) = b)\}$. When $f: A \to B$ then $range(f)$ may be B itself or any of its proper subsets.

EXERCISE 3.2.1 (WITH SOLUTION)

(a) In Exercise 3.1.1(c–e), calling the functions f, g and h respectively, identify their domain and range.

(b) Can the domain of a function ever be empty? And the range?

(c) Which of the following relations from $\mathbf{Z}$ into $\mathbf{Z}$ (see Section 1.7 to recall its definition) is a function from $\mathbf{Z}$ to $\mathbf{Z}$? Which of those that fail to be a function from $\mathbf{Z}$ to $\mathbf{Z}$ is nevertheless a partial function from $\mathbf{Z}$ to $\mathbf{Z}$?

   (i) $\{(a, |a|) : a \in \mathbf{Z}\}$

   (ii) $\{(|a|, a) : a \in \mathbf{Z}\}$

   (iii) $\{(a, a^2) : a \in \mathbf{Z}\}$

   (iv) $\{(a^2, a) : a \in \mathbf{Z}\}$

   (v) $\{(a, a+1) : a \in \mathbf{Z}\}$

(vi) $\{(a+1, a) : a \in \mathbf{Z}\}$

(vii) $\{(2a, a) : a \in \mathbf{Z}\}$

Solutions:

(a) $dom(f) = dom(g) = dom(h) = A$; $range(f) = \{1,2,3,5\}$, $range(g) = \{1,2\}$, $range(h) = \{5\}$.

(b) There is just one function with empty domain, and that is the empty function (i.e. empty relation). The empty function is also the unique function with empty range.

(c)

   (i) Yes: for every $a \in \mathbf{Z}$ there is a unique $b \in \mathbf{Z}$ with $b = |a|$.

   (ii) No: for two reasons. First, $dom(R) = \mathbf{N} \subset \mathbf{Z}$. Second, even within this domain, the 'at most one' condition is not satisfied, since a can be positive or negative.

   (iii) Yes: for every $a \in \mathbf{Z}$ there is a unique $b \in \mathbf{Z}$ with $b = a^2$.

   (iv) No: same reasons as for (ii).

   (v) Yes: for every $a \in \mathbf{Z}$ there is a unique $b \in \mathbf{Z}$ with $b = a+1$.

   (vi) Yes: every $x \in \mathbf{Z}$ is of the form $a+1$ for a unique $a \in \mathbf{Z}$, namely for $a = x-1$.

   (vii) No: $dom(R)$ is the set of all even (positive or negative) integers only.

Among those failing to be a function from $\mathbf{Z}$ to $\mathbf{Z}$, (vii) is the only one that is nevertheless a partial function from $\mathbf{Z}$ to $\mathbf{Z}$. The relations in (ii) and (iv) are not partial functions, since even within their domains the 'at most one' condition is not satisfied.

3.2.2 Image, Restriction, Closure

Let $f: A \to B$, i.e. let f be a function from A to B, and let $X \subseteq A$. The *image under f of $X \subseteq A$* is the set $\{b \in B : \exists a \in X, b = f(a)\}$, which can also be written more briefly as $\{f(a) : a \in A\}$. Thus, to take limiting cases as examples, the image $f(A)$ of A itself is $range(f)$, and the image $f(\varnothing)$ of $\varnothing$ is $\varnothing$, while the image of a singleton subset $\{a\} \subseteq A$ is the singleton $\{f(a)\}$. Thus image is not quite the same thing as value: the value of $a \in A$ under f is $f(a)$, while the image of $\{a\} \subseteq A$ under f is $\{f(a)\}$. However, many texts also use the term 'image' rather loosely as a synonym of 'value'. Always check what your author means.

When $f\colon A \to B$ and $X \subseteq A$, the image of X under f is usually written as $f(X)$, and we will follow this standard notation. However, there are contexts in which it can be ambiguous. Suppose that A is of *mixed type*, in the sense that some element X of A is also a subset of A, i.e. both $X \in A$ and $X \subseteq A$. Then the expression $f(X)$ becomes ambiguous: in a basic sense it denotes the *value* of X under f, while in a derivative sense it stands for the *image* $\{f(x) : x \in X\}$ of X under f. In this book we will rarely be discussing sets of mixed type, and so do not have to worry about the problem, but in contexts where mathematicians deal with them they sometimes avoid ambiguity by using the alternative notation $f''(X)$ for the image.

The concept of restriction is quite straightforward. Using the general definition from the theory of relations, the *restriction* of $f\colon A \to B$ to $X \subseteq A$ is the unique function on X to B that agrees with f over X. Notations vary, but one is $f|_X$. Thus $f|_X\colon X \to B$ with $f|_X(a) = f(a)$ for all $a \in X$. Because this is such a trivial operation and is rarely at the centre of attention, it is very common to cut down on distracting subscripts and use the same letter f for the function and its restriction, leaving it to context to make it clear that the domain has been reduced from A to its subset X.

Warning: When $f\colon A \to A$ and $X \subseteq A$, then the restriction of f to X will always be a function from X to A, but it will not always be a function from X to X, because there may be an $a \in X \subseteq A$ with $f(a) \in A \backslash X$. We will see a simple example of this in the next exercise.

Image should not be confused with closure. Let f be a function from A to A and let $X \subseteq A$. Then, in a 'top down' version, the *closure* $f[X]$ of X under f is the least subset of A that includes X and also includes $f(Y)$ whenever it includes Y. Equivalently, in a 'bottom up' or recursive form, we define $A_0 = X$ and $A_{n+1} = A_n \cup f(A_n)$ for each natural number n, and put $f[X] = \cup\{A_n : n \in \mathbf{N}\}$. We will make use of this notion in the chapter on recursion and induction.

EXERCISE 3.2.2 (WITH SOLUTION)

In Exercise 3.2.1(c), we saw that the relations (i) $\{(a, |a|) : a \in \mathbf{Z}\}$, (iii) $\{(a, a^2) : a \in \mathbf{Z}\}$, (v) $\{(a, a+1) : a \in \mathbf{Z}\}$, (vi) $\{(a+1, a) : a \in \mathbf{Z}\}$ are functions from $\mathbf{Z}$ to $\mathbf{Z}$. Call them *mod, square, successor, predecessor* respectively. Now recall the set $\mathbf{N} = \{0,1,2,3,\ldots\}$ from Section 1.7.

(a) Determine the image of $\mathbf{N}$ under each of these four functions.

(b) Restrict the four functions to $\mathbf{N}$. Which of them are functions into $\mathbf{N}$?

(c) Determine the images and closures of the set $A = \{-1,0,1,2\}$ under each of the four functions.

Solution:

(a) $mod(\mathbf{N}) = \{|n| : n \in \mathbf{N}\} = \{n : n \in \mathbf{N}\} = \mathbf{N}$; $square(\mathbf{N}) = \{n^2 : n \in \mathbf{N}\}$ $= \{0,1,4,9,\ldots\}$; $successor(\mathbf{N}) = \{n+1 : n \in \mathbf{N}\} = \mathbf{N}^+$; $predecessor(\mathbf{N})$ $= \{n-1 : n \in \mathbf{N}\} = \{-1\} \cup \mathbf{N}$.

(b) The only one that is not into $\mathbf{N}$ is the predecessor function, since $0-1 = -1 \notin \mathbf{N}$. It is thus a function on $\mathbf{N}$ into $\{-1\} \cup \mathbf{N}$.

(c) $Mod(A) = \{0,1,2\}$ but $mod[A] = A$; $square(A) = \{0,1,4\}$ but $square[A]$ $= \{0,1,2,4,16,32,\ldots\}$; $successor(A) = \{0,1,2,3\}$ but $successor[A] = \{-1,0,1,2,\ldots\} = \{-1\} \cup \mathbf{N}$; $predecessor(A) = \{-2,-1,0,1\}$ but $predecessor[A] = \{\ldots,-2,-1,0,1,2\} = \mathbf{Z}^- \cup \{0,1,2\}$.

Comments: (1) The restriction of $mod: \mathbf{Z} \to \mathbf{Z}$ to $\mathbf{N}$ is clearly the set of all pairs $(n,n) \in \mathbf{N}$, i.e. the function $f: \mathbf{N} \to \mathbf{N}$ that puts $f(n) = n$ for all $n \in \mathbf{N}$. This is known as the identity function over $\mathbf{N}$. For any set A, the *identity function* over A is the function $f: A \to A$ that puts $f(a) = a$ for all $a \in A$. (2) In part (c), remember that although A is not always included in its image $f(A)$, A is always included in its closure $f[A]$.

3.2.3 Composition

Perhaps the most important operation to apply to functions is that of composition. Suppose we are given two relations $R \subseteq A \times B$ and $S \subseteq B \times C$, so that the target of R is the source of S. Recall that their composition $S \circ R$ is the set of all ordered pairs (a,c) such that there is some x with both $(a,x) \in R$ and $(x,c) \in S$. Now consider the case that f is in fact a function from A to B and g is a function from B to C, briefly $f: A \to B$ and $g: B \to C$. Then $g \circ f$ is the set of all ordered pairs (a,c) such that there is some x with both $x = f(a)$ and $c = g(x)$; in other words, such that $c = g(f(a))$. It follows immediately that $g \circ f$ is a function from A to C, since for every $a \in A$ there is a unique $c \in C$ with $c = g(f(a))$.

To sum up, composition is an operation on relations which, when applied to functions with suitable domains and targets, can be defined without using the existential quantifier, and which always gives us a function. In brief notation: given functions $f: A \to B$ and $g: B \to C$ the *composition* $g \circ f: A \to C$ is a function defined by putting $g \circ f(a) = g(f(a))$. Now you can see why composition of functions and relations is written in the order we have been using: it harmonizes with standard notation for functions.

Composition is associative. Let $f: A \to B$, $g: B \to C$, $h: C \to D$. Then $(h \circ (g \circ f)) = ((h \circ g) \circ f)$ since for all $a \in A$, $(h \circ (g \circ f))(a) = h(g(f(a))) = ((h \circ g) \circ f)(a)$. In rather tedious detail: on the one hand, $(g \circ f): A \to C$ with $g \circ f(a) = g(f(a))$, so $h \circ (g \circ f): A \to D$ with

$(h \circ (g \circ f))(a) = h(g(f(a)))$. On the other hand, $(h \circ g): B \to C$ with $(h \circ g)(b) = h(g(b))$, so $(h \circ g) \circ f: A \to D$ with also $((h \circ g) \circ f)(a) = h(g(f(a)))$.

However, composition of functions is not in general commutative. Let $f: A \to B$, $g: B \to C$. Then the composition $g \circ f$ is a function, but $f \circ g$ is not a function unless also $C = A$. Even in the case that $A = B = C$, so that $g \circ f$ and $f \circ g$ are both functions, they may not be the same function. For example, suppose that $A = B = C = \{1,2\}$, with $f(1) = f(2) = 1$ while $g(1) = 2$ and $g(2) = 1$. Then $g \circ f(1) = g(f(1)) = g(1) = 2$, while $f \circ g(1) = f(g(1)) = f(2) = 1$.

EXERCISE 3.2.3

    (a) Draw a diagram to illustrate the above example of non-commutativity of composition.

    (b) Give an example of functions $f,g: \mathbf{N} \to \mathbf{N}$ such that $f \circ g \neq g \circ f$.

    (c) Show that we can generalize the definition of composition a little, in the sense that the composition $g \circ f: A \to C$ is a function whenever $f: A \to B$ and $g: B' \to C$ and $B \subseteq B'$. Give an example to show that $g \circ f$ may not be a function when, conversely, $B' \subseteq B$.

3.2.4 Inverse

We recall the notion of the *converse* (alias *inverse*) R^{-1} of a relation: it is the set of all ordered pairs (b,a) such that $(a,b) \in R$. The converse of a relation R is thus always a relation. But when R is a function, the converse *is sometimes a function, sometimes not.*

EXERCISE 3.2.4 (WITH PARTIAL SOLUTION)

    (a) Let $A = \{1,2,3\}$ and $B = \{a,b,c,d\}$. Let $f = \{(1,a), (2,a), (3,b)\}$ and $g = \{(1,a), (2,b), (3,c)\}$. (i) Explain why neither $f^{-1} = \{(a,1), (a,2), (b,3)\}$ nor $g^{-1} = \{(a,1), (b,2), (c,3)\}$ is a function from B to A. (ii) Draw a diagram of the example.

    (b) Give examples of (i) a function $f: \mathbf{Z} \to \mathbf{Z}$ whose inverse is not a function, (ii) a function $g: \mathbf{N} \to \mathbf{N}$ whose inverse is not a function.

Partial solution to (a): f^{-1} is not a function at all, because we have both $(a,1), (a,2)$ in it. On the other hand, g^{-1} is a function, but not from B to A, but rather from the proper subset $B' = \{a,b,c\} \subset B$ to A. It is thus a partial function from B to A.

The existence of an inverse function is closely linked with two further concepts, injectivity and surjectivity, which we now examine.

3.3 Injections, Surjections, Bijections

3.3.1 Injectivity

Let $f: A \to B$. We say that f is *injective* (alias one-one) iff whenever $a \neq a'$ then $f(a) \neq f(a')$. In words, iff it takes distinct arguments to distinct values (distinct inputs to distinct outputs). Contrapositively: iff whenever $f(a) = f(a')$ then $a = a'$. In other words, iff for each $b \in B$ there is at most one $a \in A$ with $f(a) = b$.

Of the two functions f,g described in Exercise 3.2.4, f is not injective, since $f(1) = f(2)$ although $1 \neq 2$. However, g is injective since it takes distinct arguments to distinct values.

EXERCISE 3.3.1 (WITH PARTIAL SOLUTION)

(a) In an earlier exercise, we saw that the relations (i) $\{(a, |a|) : a \in \mathbf{Z}\}$, (iii) $\{(a, a^2) : a \in \mathbf{Z}\}$, (v) $\{(a, a+1) : a \in \mathbf{Z}\}$, (vi) $\{(a+1, a) : a \in \mathbf{Z}\}$ are functions over $\mathbf{Z}$, i.e. from $\mathbf{Z}$ to $\mathbf{Z}$, called *mod, square, successor, predecessor* respectively. Which of them are injective?

(b) What does injectivity mean in terms a digraph for the function?

(c) What does injectivity mean in terms a table for the function?

(d) Show that the composition of two injective functions is injective, and give examples to show that the failure of injectivity in either of the two components can lead to failure of injectivity for the composition.

Solution to (a): Over $\mathbf{Z}$, *mod* is not injective, since e.g. $|1| = |-1|$, likewise for *square* since e.g. $3^2 = (-3)^2$. On the other hand, *successor* is injective, since $a+1 = a'+1$ implies $a = a'$ for all $a,a' \in \mathbf{Z}$. Similarly, *predecessor* is injective, since $a = a'$ implies $a+1 = a'+1$.

Comment: This exercise brings out the importance of always keeping clearly in mind the intended domain of the function when checking whether it is injective. Take for example the function of squaring. As we have just seen, when understood with domain $\mathbf{Z}$, it is not injective. But when understood with domain $\mathbf{N}$ (i.e. as the restriction of the former function to $\mathbf{N}$) then it is injective, since for all natural numbers $a,b, a^2 = b^2$ implies $a = b$.

Note that the injectivity of a function from A to B is not enough to guarantee that its inverse is a function from B to A. For example, as noted in the comments on the last exercise, the function of squaring on $\mathbf{N}$ (rather than on $\mathbf{Z}$) is injective. But its inverse is not a function on $\mathbf{N}$, since there are elements of $\mathbf{N}$, e.g. 5, that are not in its domain, not being the square of any natural number.

Nevertheless, injectivity gets us part of the way: it suffices to ensure that the inverse relation f^{-1} of a function from A to B is a function from $range(f) \subseteq B$ to A, and so is a *partial function* from B to A. Reason: From our work on relations, we know that f^{-1} must be a *relation* from $range(f)$ to A, so we need only show that for all $b \in B$ there is at most one $a \in A$ such that $(b,a) \in f^{-1}$, i.e. at most one $a \in A$ with $(a,b) \in f$. But this is exactly what is given by the injectivity of f. Indeed, we can also argue in the converse direction, with the result that we have the following equivalence: *a function $f: A \rightarrow B$ is injective iff its inverse relation f^{-1} is a function from range(f) to A.*

3.3.2 Surjectivity

Let $f: A \rightarrow B$. We say that f is *onto B* or *surjective* (with respect to B) iff for all $b \in B$ there is some $a \in A$ with $f(a) = b$. In other words, iff every element of B is the value of some element of A under f. Equivalently, iff $range(f) = B$.

For example, if $A = \{1,2,3\}$ and $B = \{7,8,9\}$ then the function f that puts $f(1) = 9$, $f(2) = 8$, $f(3) = 7$ is onto B, but is not onto $B' = \{6,7,8,9\}$, since it 'misses out' the element $6 \in B'$.

EXERCISE 3.3.2

(a) What does surjectivity mean in terms a digraph for the function?

(b) What does it mean in terms a table for the function?

For some more substantive examples, we look to the number systems. Over $\mathbf{Z}$ both of the functions $f(a) = a+1$ and $g(a) = 2a$ are injective, but only f is onto $\mathbf{Z}$, since the odd integers are not in the range of g. However, if we restrict these two functions to the set $\mathbf{N}$ of natural numbers, then not even f is onto $\mathbf{N}$, since $0 \in \mathbf{N}$ is not the successor of any natural number.

EXERCISE 3.3.3 (WITH PARTIAL SOLUTION)

(a) Consider the function $f(a) = a^2$ over $\mathbf{N}$, $\mathbf{N}^+$, $\mathbf{Z}$, $\mathbf{Q}$, $\mathbf{R}$ respectively and determine in each case whether it is surjective.

(b) Show that the composition of two surjective functions is surjective, and give examples to show that the failure of surjectivity in either of the two components can lead to its failure for the composition.

Solution to (a): For $\mathbf{N}$, $\mathbf{N}^+$, $\mathbf{Z}$, $\mathbf{Q}$ the answer is negative, since e.g. 2 is in each of these sets, but there is no number a in any of these sets with $a^2 = 2$. In common parlance, 2 does not have a rational square root. For $\mathbf{R}$, the answer is positive, since 2 does have a real square root: for any real b there is an $a \in \mathbf{R}$ with $a^2 = b$.

Both of the terms 'onto' and 'surjective' are in common use, but the former is more explicit in that it makes it easier for us to say onto *what*. We can say simply 'onto B', whereas it is rather more longwinded to say 'surjective with respect to B'. Whichever of the two terms one uses, it is important to be clear what set B is intended, since quite trivially *every* function is onto *some* set – namely its range!

Alice Box: Terminology for functions

Alice: I'm getting a bit tired of all these terminological variations! Isn't it time that you people got together and agreed on a common way of speaking?

Hatter: In mitigation, remember that the concept of function has a very long history across a number of different mathematical communities. The terminological variants are often traces of that evolution. For example, in this text we are using the terms 'converse' and 'inverse' interchangeably. But the former term was preferred by those focussed on the theory of relations, while the latter was favoured by those looking primarily at functions. Indeed, for a long time, function people tended to ignore non-functional relations altogether. As a result they phrased some matters rather differently saying, for instance, that the inverse of a function *does not exist* (whereas we would say, exists as a relation but not *as a function*) unless it is injective.

Alice: Traps at every corner!

Hatter: Indeed. Half the trouble that students have with simple mathematics is due to quirks in the language in which it is conveyed to them. Here is another example. There is a certain ambiguity when we use the expression $f(x)$, when f is a function from A to B. Normally, it stands for the unique $y \in B$ with

(Continued)

Alice Box: (Continued)

$(x,y) \in f$, i.e. for the value of the function f for argument x. But sometimes it stands for the function f itself, which is a set of ordered pairs, with the x merely serving to remind us what letters will be employed to refer to the arguments. If you read back over this chapter, you will find that we have used the expression both ways – in the former way in the theory, and the latter in some of the examples, where it helps us to be brief.

3.3.3 Bijective Functions

A function $f: A \to B$ that is both injective and onto B is said to be a *bijection* between A and B. An older term sometimes used is *one-one correspondence*. An important fact about bijectivity is that it is equivalent to the inverse being a function from B. That is, *a function $f: A \to B$ is a bijection between A and B iff its inverse f^{-1} is a function from B to A.*

Proof By definition, f is a bijection between A and B iff f is injective and onto B. As we saw earlier in the chapter, $f: A \to B$ is injective iff f^{-1} is a partial function from B to A; and moreover f is onto B iff $range(f) = B$, i.e. iff $dom(f^{-1}) = B$. Putting this together, f is a bijection between A and B iff f^{-1} is a partial function from B to A with $dom(f^{-1}) = B$, i.e. iff f^{-1} is a function from B to A.

Alice box : Proving an equivalence

Alice: Why didn't you prove this equivalence in the same way as before, breaking it into two parts, one the converse of the other, and proving each separately by conditional proof?

Hatter: We could perfectly well have done that. But in this particular case the argument going in one direction would have turned out to be essentially the same as its converse, run backwards. So in this case, we can economize by doing it all as a chain of *iffs*. However, such an economy is not always available.

EXERCISE 3.3.4

(a) What is a bijection in terms a (i) digraph, (ii) a table for the function?

(b) Show that the inverse of a bijection from A to B is a bijection from B to A. *Hint*: you know that it is a function from B to A, so you only need check out injectivity and surjectivity.

We end this section with a final remark on inverses. We have seen that the inverse of a function $f: A \rightarrow B$ is a relation from $f(A) \subseteq B$ back to A, but will not be a function unless f is injective. However, if we rise one level of abstraction and take the value of f^{-1}, for a given $b \in f(A) \subseteq B$, to be *the set of all $a \in A$ such that $f(a) = b$*, then this is *always* a function on $f(A)$ but into the power set $\mathcal{P}(A)$ of A. With a little abuse of notation, we can write it briefly, with the same notation, as $f^{-1}: f(A) \rightarrow \mathcal{P}(A)$. An example of this will arise in the chapter on probability.

3.4 Using Functions to Compare Size

One of the many uses of functions is to compare the sizes of sets. In this section, we will consider only finite sets, although a more advanced treatment could also consider infinite ones. Recall from Chapter 1 that when A is a finite set, we write $\#(A)$ for the number of elements of A, also known as the *cardinality* of A. Another common notation for this is $|A|$. We will look at two principles, one for bijections and one for injections.

3.4.1 The Equinumerosity Principle

Let A, B be finite sets. The *equinumerosity principle* says : $\#(A) = \#(B)$ *iff there is some bijection $f: A \rightarrow B$.*

Proof Suppose first that $\#(A) = \#(B)$. Since both sets are finite, there is some natural number n with $\#(A) = n = \#(B)$. Let $a_1, \ldots, a_n$ be the elements of A, and $b_1, \ldots, b_n$ those of B. Let $f: A \rightarrow B$ be the function that puts each $f(a_i) = b_i$. Clearly f is injective and onto B. For the converse, let $f: A \rightarrow B$ be a bijection. Suppose A has n elements $a_1, \ldots, a_n$. Then the list $f(a_1), \ldots, f(a_n)$ enumerates all the elements of B, counting none of them twice, so B also has n elements.

A typical application of this principle takes the following form, using only one half of the *iff*. We have two sets A, B and want to show that they have the same cardinality. We look for some function that is a bijection between the two. If we find one, then we are done.

EXERCISE 3.4.1 (WITH SOLUTION)

(a) Let A be the set of sides of a polygon, and B the set of its vertices. Show $\#(A) = \#(B)$.

(b) In a reception, everyone shakes hands with everyone else just once. If there are n people in the reception, how many handshakes are there?

Solution:

(a) Let $f\colon A \to B$ be defined by associating each side with its right endpoint. Clearly this is both injective and is onto B. Hence $\#(A) = \#(B)$.

(b) Let A be the set of the handshakes, and let B be the set of all unordered pairs $\{x,y\}$ of distinct people (i.e. $x \neq y$) from the reception. Define $f\colon A \to B$ by associating each handshake with the two distinct people involved in it. Clearly this gives us a bijection between A and B. So we need only ask how many elements there are in B. Clearly there are $n \cdot (n{-}1)$ ordered pairs of distinct people in the reception, and thus $n \cdot (n{-}1)/2$ unordered pairs, since each unordered pair $\{x,y\}$ with $x \neq y$ corresponds to two distinct ordered pairs (x,y) and (y,x). Thus there are $n \cdot (n{-}1)/2$ handshakes.

As these examples reveal, the sought-for bijection can be a quite simple and obvious one, and often it associates the elements of some rather 'everyday' set (like the set of all handshakes) with some rather more 'abstract' one (like the set of all unordered pairs $\{x,y\}$ of distinct people from the reception). The bijection allows us to apply to the everyday set the counting rules that we know for the abstract one.

3.4.2 The Principle of Comparison

Let A,B be finite sets. The *principle of comparison* says: $\#(A) \leq \#(B)$ *iff there is some injective function* $f\colon A \to B$.

Proof We use the same sort of argument as for the equinumerosity principle. Suppose first that $\#(A) \leq \#(B)$. Since both sets are finite, there is some $n \geq 0$ with $\#(A) = n$, so that $\#(B) = n{+}m$ for some $m \geq 0$. Let $a_1, \ldots, a_n$ be the elements of A, and $b_1, \ldots, b_n, \ldots, b_{n+m}$ those of B. Let $f\colon A \to B$ be the function that puts each $f(a_i) = b_i$. Clearly f is injective (but not necessarily onto) B. For the converse, let $f\colon A \to B$ be injective. Suppose A has n elements $a_1, \ldots, a_n$. Then the list $f(a_1), \ldots, f(a_n)$ enumerates some (but not necessarily all) the elements of B, counting none of them twice, so B has at least n elements, i.e. $\#(A) \leq \#(B)$.

Alice Box: Equinumerosity for infinite sets

Alice: You formulated the equinumerosity principle and the principle of comparison in terms of finite sets, saying cryptically they could also be formulated for infinite ones. How would you do that?

(Continued)

Alice Box: (Continued)

Hatter: The first step is to decide what you mean by saying that one set has the same size or cardinality as another. In the finite case we understand that notion before we ever hear of bijections, but when the set is infinite we need to explain what it means. This was done by Georg Cantor towards the end of the nineteenth century, who simply *defined* it in terms of bijections: sets A and B, finite or infinite, are defined to have the same cardinality iff there is a bijection from one to the other. So the equinumerosity principle becomes true by definition.

Alice: Doesn't that give all infinite sets the same cardinality? Isn't there a bijection between any two infinite sets?

Hatter: At first sight it might seem so, and indeed the sets $\mathbf{N}$, $\mathbf{N}^+$, $\mathbf{Z}$ and $\mathbf{Q}$ are all equinumerous in Cantor's sense: it is possible to find a bijection between any two of them. Any set that has a bijection with $\mathbf{N}$ is said to be *countable*, so $\mathbf{N}^+$, $\mathbf{Z}$ and $\mathbf{Q}$ are countable. But one of Cantor's fundamental theorems was to show, surprisingly, that this is not the case for $\mathbf{R}$: it is not countable – there is no bijection between it and $\mathbf{N}$. Further, he showed that there is no bijection between any set A and its power set $\mathcal{P}(A)$.

Alice: How can you prove this?

Hatter: The proof is ingenious, and short. It uses what is known as the 'diagonal construction'. But it would take us out of our main path. You will find it in any good introduction to set theory for students of mathematics, e.g. Paul Halmos' *Naïve Set Theory*.

3.4.3 The Pigeonhole Principle

In applications of the principle of comparison, we typically use only the right-to-left half of it, formulating it contrapositively as follows. Let A,B be finite sets. If $\#(A) > \#(B)$ then no function $f\colon A \to B$ is injective. In other words, *if* $\#(A) > \#(B)$ *then for every function* $f\colon A \to B$ *there is a* $b \in B$ *such that* $b = f(a)$ *for two distinct* $a \in A$.

This simple rule is known as the *pigeonhole principle*, from the example in which A is the (large) set of letters to be delivered to people in the office, and B is the (small) set of their pigeonholes. It also has a more general form, as follows. *Let* A,B *be finite sets. If* $\#(A) > k\cdot\#(B)$ *then for every function* $f\colon A \to B$ *there is a* $b \in B$ *such that* $b = f(a)$ *for at least* $k+1$ *distinct* $a \in A$.

The pigeonhole principle is surprisingly versatile as a way of showing that at in suitable situations, at least two distinct items must have a certain property. The wealth of possible applications can only be appreciated by looking at examples. We begin with a very straightforward one.

EXERCISE 3.4.2 (WITH SOLUTION)

A village has 400 inhabitants. Show that at least two of them have the same birthday, and that at least 34 are born in the same month of the year.

Solution: Let A be the set of people in the village, B the set of the days of the year, C the set of months of the year. Let $f\colon A \to B$ associate with each villager his or her birthday, while $g\colon A \to C$ associates with the month of birth. Since $\#(A) = 400 > 366 \geq \#(B)$ the pigeonhole principle tells us that there is a $b \in B$ such that $b = f(a)$ for two distinct $a \in A$. This answers the first part of the question. Also, since $\#(A) = 400 > 396 = 33 \cdot \#(C)$, the generalized pigeonhole principle tells us that there is a $c \in C$ such that $c = f(a)$ for $33+1 = 34$ distinct $a \in A$, which answers the second part of the question.

In this exercise, it was quite obvious what sets A, B and what function $f\colon A \to B$ to choose. In other examples, this may need more reflection, and sometimes considerable ingenuity.

EXERCISE 3.4.3 (WITH SOLUTION)

In a club, everyone has just one given name and just one family name. It is decided to refer to everyone by two initials, the first initial being the first letter of the given name, the second initial being the first letter of the family name. How many members must the club have to make it inevitable that two distinct members are referred to by the same initials?

Solution: It is pretty obvious that we should choose A to be the set of members of the club. Let B be the set of all ordered pairs of letters from the alphabet. The function $f\colon A \to B$ associates with each member a pair as specified in the club decision. Assuming that we are dealing with a standard English alphabet of 26 letters, $\#(B) = 26.26 = 676$. So if the club has 677 members, the pigeonhole principle guarantees that $f(a) = f(a')$ for two distinct $a, a' \in A$.

When solving problems by the pigeonhole principle, always begin by choosing carefully the sets A, B and the function $f: A \to B$. Indeed, this will often be the key step in finding the solution. Sometimes, as in the above example, the elements of B will be rather abstract items, such as ordered n-tuples.

EXERCISE 3.4.4

A multiple-choice exam has 5 questions, each with two possible answers. Assuming that each student enters a cross in exactly one box of each question (no unanswered, over-answered, or spoiled papers), how many students need to be in the class to guarantee that at least four students submit the same answer paper?

3.5 Some Handy Functions

In this section we look at some very simple functions that appear over and again: identity, constant, projection and characteristic functions, as well as the deployment of functions to represent sequences and families. The student may find it difficult to assimilate all these in one sitting. No matter, they are here to be recalled whenever needed at a later moment.

3.5.1 Identity Functions

Let A be any set. The identity function on A is the function $f: A \to A$ such that for all $a \in A$, $f(a) = a$. As simple as that! It is sometimes written as i_A, or with the Greek letter iota as ι_A. It is very important in abstract algebra, and comes up often in computer science, never at the centre of attention but (like the restriction of a function) as an everyday tool used almost without thinking.

EXERCISE 3.5.1

(a) Show that the identity function over any set A is a bijection, and is its own inverse.

(b) Let $f: A \to B$. Show that $f \circ i_A = f = i_B \circ f$.

Alice Box: One identity function or many?

Alice: So, for every choice of set A you get a different identity function i_A?

Hatter: Strictly speaking, yes.

Alice: Why not define a great big identity function, for once and for all, by putting $i(a) = a$ for every a whatsoever?

Hatter: Attractive, but there is a technical problem. What would its domain and range be?

Alice: The universal set, i.e. the set of all things whatsoever.

Hatter: Unfortunately, as we saw in Chapter 1.4, standard set theory as understood today does not admit such a set, on pain of contradiction. So we must relativize the concept to whatever 'local universe' set U we are working in. A little bit of a bother, but not too bad in practice.

3.5.2 Constant Functions

Let A, B be non-empty sets. A *constant function* on A into B is any function $f: A \to B$ such that for some $b \in B$ we have $f(a) = b$ for all $a \in A$.

The order of the quantifiers is vital here: we require that $\exists b \in B \, \forall a \in A \, f(a) = b$, in other words that all the elements of A have the same value under f. This is much stronger than requiring merely that $\forall a \in A \, \exists b \in B \, f(a) = b$; in fact, *that* holds for any function f whatsoever! In general, there is an immense difference between statements of the form $\forall x \exists y (\ldots)$ and corresponding ones of the form $\exists x \forall y (\ldots)$. In a later chapter on the logic of the quantifiers, we will set out systematically the relations between the various propositions $Q x Q' y (\ldots)$, where Q and Q' are quantifiers (universal or existential), x and y are their attached variables, and the expression $(\ldots)$ is kept fixed.

EXERCISE 3.5.2

(a) Fix non-empty sets A, B with $\#(A) = m$ and $\#(B) = n$. How many constant functions $f: A \to B$ are there?

(b) Show that when $\#(A) > 1$ then no constant function $f: A \to B$ is injective, and when $\#(B) > 1$ then no constant function $f: A \to B$ is onto B.

(c) Let $f: A \to B$ and $g: B \to C$. Show that if either of f, g is a constant function, then $g \circ f: A \to C$ is a constant function.

3.5.3 Projection Functions

Let $f\colon A \times B \to C$ be a function of two arguments, and let $a \in A$. By the *right projection of f at a* we mean the one-argument function $f_a\colon B \to C$ defined by putting $f_a(b) = f(a,b)$ for each $b \in B$.

Likewise, letting $b \in B$, the *left projection of f at b* is the one-argument function $f_b\colon A \to C$ defined by putting $f_b(a) = f(a,b)$ for each $a \in A$.

In other words, to form the left or right projection of a (two-argument) function, we hold one of the arguments of f fixed at some value, and consider the (one-argument) function obtained by allowing the other argument to vary.

3.5.4 Characteristic Functions

Let U be any set fixed as a local universe. For each subset $A \subseteq U$ we can define a function $f_A\colon U \to \{1,0\}$ by putting $f_A(u) = 1$ when $u \in A$ and $f_A(u) = 0$ when $u \notin A$. This is known as the *characteristic function* of A (modulo the universe U). Thus the characteristic function f_A specifies the truth-value of the statement that $u \in A$.

Conversely, when $f\colon U \to \{1.0\}$, we can define the *associated subset* of U by putting $A_f = \{u \in U : f(a) = 1\}$.

Clearly, there is a bijection between the subsets of U and the functions $f\colon U \to \{1.0\}$, and in fact we can make either do the work of the other. In some contexts, it is notationally more convenient to work with characteristic functions rather than subsets.

3.5.5 Families of Sets

In Section 1.5, we introduced sets of sets. They are usually written $\{A_i : i \in I\}$ where the A_i are sets and the set I, called an *index set*, helps us keep track of them. Because the phrase 'set of sets' tends to be difficult for the mind to process, we also speak of $\{A_i : i \in I\}$ as a *collection* of the sets A_i; but the term 'collection' does not mean anything new – it is merely to facilitate reading.

We now introduce the subtly different concept of a *family of sets*. This refers to any *function* on a domain I (called an *index set*) such that for each $i \in I$, $f(i)$ is a set. Writing $f(i)$ as A_i, it is thus the set of all ordered pairs $(i, f(i)) = (i, A_i)$ with $i \in I$. The range of this function is the collection $\{A_i : i \in I\}$.

The difference is subtle, and in some contexts sloppiness does not matter. But in certain situations, notably applications of the pigeonhole principle and other counting rules that we will come to in a later chapter, it is very important. It can happen, for example, that the index set I has n elements, say $I = \{1, \ldots n\}$, but the function f is not injective, say $f(1) = f(2)$, i.e. $A_1 = A_2$. In that case the *family*, containing all the pairs (i, A_i) with $i \leq n$, has n elements, but the *collection*, containing all the sets A_i with $i \leq n$, has fewer elements since $A_1 = A_2$.

Once more, the substance of mathematics is intricately entwined with the way that it uses language. The convention of subscripting, wherever it occurs, is in effect an implicit way of describing a function. For example, when we say 'let p_i be the ith prime number, for any $i \in \mathbf{N}^+$', then we are implicitly considering the function $p\colon \mathbf{N}^+ \to \mathbf{N}^+$ such that each $p(i)$ is the ith prime number. The notation with subscripts can often be easier to read than the standard function notation, as it gets rid of brackets. And in some contexts, all we really need to know about the function is its range, so *in those contexts* we can treat it as merely a collection of sets.

3.5.6 Sequences

The uses of functions are endless. In fact, their role is so pervasive that some mathematicians prefer to see them, rather than sets, as providing the bedrock of their discipline. We will not venture into this question, which belongs rather to the philosophy of mathematics, but instead illustrate the versatility of functions by seeing how they can clarify the notion of a *sequence*.

In computer science as in mathematics itself, we often need to consider sequences $a_1, a_2, a_3, \ldots$ of items. The items a_i might be numbers, sets or other mathematical objects; in computer science they may be the instructions in a program, or steps in its execution. The sequence itself may be finite, with just n terms $a_1, \ldots, a_n$ for some natural number n, or infinite with a term a_i for each positive integer i, in which case we usually write it in an informal suspended dots notation, as $a_1, a_2, a_3, \ldots$. But what is such a sequence?

It is convenient to identify an infinite sequence $a_1, a_2, a_3, \ldots$ with a function $f\colon \mathbf{N}^+ \to A$ for some appropriately chosen set A, with $f(i) = a_i$ for each $i \in \mathbf{N}^+$. The ith term in the sequence is thus just the value of the function for argument (input) i.

When the sequence is finite, there are two ways to go. We can continue to identify it with a function $f\colon \mathbf{N}^+ \to A$ with $f(i) = a_i$ for each $i \leq n \in \mathbf{N}^+$ and with $f(n+j) = f(n)$ for all $j \in \mathbf{N}^+$, so that the function becomes constant in its value from $f(n)$ upwards. Or, more intuitively, we can take it to be a function $f\colon \{i \leq n \in \mathbf{N}^+\} \to A$, i.e. as a partial function on $\mathbf{N}^+$ with domain $\{i \leq n \in \mathbf{N}^+\}$. It doesn't matter really which way we do it; in either case we have made a rather vague notion sharper by explaining it in terms of functions.

Alice Box: n-tuples, sequences, strings and lists

Alice: I'm getting confused. So a finite *sequence* of elements of A is a function $f: \{i \leq n \in \mathbf{N}^+\} \to A$ and you write it as $a_1, a_2, \ldots, a_n$. But what's the difference between this and an ordered *n-tuple* $(a_1, a_2, \ldots, a_n)$? While we are at it, I have been reading around, and I also find people talking about finite *strings* and finite *lists*. Are they all the same, or different?

Hatter: Well, er...

Alice: They come at me with different notations. Tuples and sequences are written as above, but strings are written just as $a_1 a_2 \ldots a_n$ with neither commas nor external brackets. Lists are written $<a_1, a_2, \ldots, a_n>$ or sometimes with further internal brackets, say $<a_1 <a_2, \ldots, a_n>>$. And the symbols used for the empty tuple, sequence, string and list are all different. Help! What is going on?

Hatter: Indeed, this is quite confusing, not to use a more impolite term. Let me try to sort it out for you, in a rather informal manner. At bottom, these are all the same kind of object, but built up in different ways and with different tools for their manipulation. The most abstract concept is that of an *n*-tuple. We don't care what it is; all we care about is the criterion for identity that we mentioned in the previous chapter: $(x_1, x_2, \ldots, x_n) = (y_1, y_2, \ldots, y_n)$ iff $x_i = y_i$ for all i from 1 to n. An *n*-tuple can be *anything that satisfies that condition*.

Alice: And sequences?

Hatter: They are more specific. As we have said, a sequence is a *function* $f: \{i \leq n \in \mathbf{N}^+\} \to A$. Sequences happen to satisfy the identity criterion that I mentioned, and so we may regard them as a particular kind of tuple. Mathematicians like them because they are accustomed to working with functions on $\mathbf{N}^+$.

Alice: Strings?

Hatter: Think of them as tuples, usually of symbols, that come equipped with a tool for putting them together and taking them apart. This tool is the operation of *concatenation*, which consists of taking any strings s and t and forming a string $con(s, t)$. In the case of symbols, this is the longer symbol formed by writing s and then immediately to its right the other component symbol t. When we talk of strings, we are thinking of tuples where concatenation is the only available way, or the only allowed one, of building or dismantling them. Computer scientists like them, as concatenation is an operation that they are familiar with.

Alice: And lists?

Hatter: They too are tuples, again often of symbols, and again equipped with a single tool for their construction and decomposition. But this time the tool

(Continued)

Alice Box: (Continued)

is different, and more limited in its power. It can be thought of as a restricted form of concatenation. We are allowed to take any list y and put in front of it an element a of the base set A of elementary symbols (often called the *alphabet* for constructing lists). This forms the slightly longer list $<a,y>$. Here a is called the *head* of the list, and y is the *tail* of the list $<a,y>$. If y itself is compound, say $y = <b,x>$ where $b \in A$ and x is a shorter list, then $<a,y> = <a, <b,x>>$, and so on. Being a restricted form of concatenation, the operation is given the very similar name *cons*, so that $cons(a,y) = <a,y>$. The important thing about the *cons* operation is that while it can take any list as its second argument, it can take only elements of the alphabet A as its first argument. It is in effect a restriction of the first argument of the concatenation operation to the alphabet set A.

Alice: Why this restriction?

Hatter: One reason is that is satisfies a special mathematical condition of unique decomposability, which concatenation, for example, does not satisfy. This permits us to carry out definitions by structural recursion, as we will explain in Chapter 4. Another reason is that it is particularly easy to implement on a computer.

Alice: So the basic difference between tuples, sequences, strings and lists is not so much what they *are* but what *tools they come with*, to build and manipulate them?

Hatter: That's quite a good way to put it! However, after making these fine distinctions, I must warn you yet again that authors can be rather loose in the way they write. And, just to annoy you further, strings are often also called *words*.

FURTHER EXERCISES

3.1. *Partial functions*

(a) Characterize the notion of a partial function from A to B in terms of (i) its table and (ii) its digraph as a relation.

(b) Let R be a relation from A to B. Show that it is a partial function from A to B iff it is a function from $dom(R)$ to B.

3.2. *Image, closure*

(a) The *floor function* from $\mathbf{R}$ into $\mathbf{N}$ is defined by putting $\lfloor x \rfloor$ to be the largest integer less than or equal to x. What are the images under the floor function of the sets $[0,1] = \{x \in \mathbf{R} : 0 \le x \le 1\}$, $[0,1) = \{x \in \mathbf{R} : 0 \le x < 1\}$, $(0,1] = \{x \in \mathbf{R} : 0 < x \le 1\}$, $(0,1) = \{x \in \mathbf{R} : 0 < x < 1\}$?

(b) Show that when $f(A) \subseteq A$ then $f[A] = A$.

(c) Let $f\colon A \to B$ be a function from set A into set B. Recall that for each $b \in B, f^{-1}(b)$ is the set $\{a \in A\colon f(a) = b\}$. (i) Show that the family of all these sets $f^{-1}(b)$ for $b \in f(A) \subseteq B$ is a partition of A in the sense defined in Chapter 2. (ii) Is this still the case if we include in the family the sets $f^{-1}(b)$ for $b \in B \setminus f(A)$?

(d) Show that for any partition of A, the function f taking each element $a \in A$ to its cell is a function on A into the power set $P(A)$ of A with the partition as its range.

3.3. *Injections, surjections, bijections*

(a) Is the floor function from $\mathbf{R}$ into $\mathbf{N}$ injective? (ii) Is it onto $\mathbf{N}$?

(b) Use the results of exercises in this chapter to show that the composition of two bijections is a bijection.

(c) Use the equinumerosity principle to show that there is never any bijection between a finite set and any of its proper subsets.

(d) Give an example to show that there can be a bijection between an infinite set and certain of its proper subsets.

(e) Let A,B be sets with B finite. Show that when $f\colon A \to B$ is onto B (but not necessarily injective), then there is an injective function $g\colon B \to A$ with $g(b) \in f^{-1}(b)$ for all $b \in B$. *Hint*: To get the idea, draw a picture of a simple example.

(f) Use the principle of comparison to show that for finite sets A,B, if there are injective functions $f\colon A \to B$ and $g\colon B \to A$, then there is a bijection from A to B.

3.4. *Pigeonhole principle*

(a) Use the general form of the pigeonhole principle to show that of any seven propositions, there are at least four with the same truth-value.

(b) Let $K = \{n \in \mathbf{N}^+ : n \leq 16\}$. How many distinct numbers must be selected from K to guarantee that the sum of two of them is 18? *Hint*: Let B be the set of all unordered pairs $\{x,y\}$ with $x,y \in K$ and $x+y = 18$.

3.5. *Handy functions*

(a) Describe the characteristic function, for the local universe of positive integers, of (i) the set of all odd numbers, (ii) the set of all prime numbers.

(b) Describe the right projection of the multiplication function $x \cdot y$ at the value $x = 3$. Also the left projection at the value $y = 3$. Are they the same function? Why?

Selected Reading

Paul R. Halmos *Naive Set Theory*. Springer, 2001 (new edition), Chapters 8, 10.

Seymour Lipschutz *Set Theory and Related Topics*. McGraw Hill Schaum's Outline Series, 1998, Chapters 4–5.

James L. Hein *Discrete Structures, Logic and Computability*. Jones and Bartlett, 2002 (second edition) Chapter 2.

4

Recycling Outputs as Inputs: Induction and Recursion

Chapter Outline

This chapter introduces induction and recursion, which are omnipresent in computer science.

The simplest context in which they arise is in the domain of the positive integers, and that is where we begin. We explain induction as a method for *proving facts about the positive integers*, and recursion as a method for *defining functions* on the same domain. We will also distinguish two different methods for *evaluating* such functions.

From this familiar terrain, the basic concepts of recursion and induction can be extended to structures, processes and procedures of many kinds, not only numerical ones. Particularly useful for computer scientists are the forms known as *structural induction and recursion*, and we give them special attention. We will look at structural recursion as a way of *defining sets*, structural induction as a way of *proving* things about those sets, and then structural recursion once more as a way of *defining functions with recursively defined domains*. At this last point special care is needed, as the definitions of such functions succeed only when a special condition of *unique decomposition* is satisfied. Happily, it holds in many computer science applications.

The broadest and most powerful kind of induction/recursion may be formulated for sets of any kind, provided only they are equipped with a relation that is *well-founded*, in a sense we explain. All other kinds may be seen as special cases of that one.

D. Makinson, *Sets, Logic and Maths for Computing*,
DOI: 10.1007/978-1-84628-845-6_4, © Springer-Verlag London Limited 2008

In a final section we look at the notion of a recursive *program*, and see how the ideas that we have developed in the chapter are manifested there.

4.1 What are Induction and Recursion?

The two words are used in different contexts. 'Induction' is the term more commonly applied when talking about *proofs*. 'Recursion' is the one used in connection with *definitions and constructions*. We will follow this tendency, speaking of inductive proofs and recursive definitions. But it should not be thought that they answer to two fundamentally different topics: the same basic idea is involved in each.

What is this basic idea? It will help if we look for a moment at the historical context. We are considering an insight that goes back to ancient Greece and India, but whose explicit articulation had difficulty breaking free from a long-standing rigidity. From the time of Aristotle onwards it was a basic tenet of logic, and of science in general, that nothing should ever be defined in terms of itself, on pain of making the definition circular. Nor should any proof assume what it is setting out to prove, for that too would create circularity.

Taken strictly, these precepts remain perfectly true. But it has been realized that definitions, proofs, and procedures may also 'call upon' themselves, in the sense that later steps may systematically appeal to the outcome of earlier steps. In suitable contexts, the value of a function for a given value of its argument may be defined in terms of its value for smaller arguments; a proof of a fact about an item may assume that we have already proven the same fact about earlier items; an instruction telling us how to carry out steps of a procedure or program may specify this in terms of what the previous steps have already done. In each case, what we need is a *clear stepwise ordering of the domain we are working on, with a clearly specified starting point.*

What do these requirements mean? To clarify them we begin by looking at the simplest context in which they are satisfied: the positive integers. There we have a definite starting point, 1. We also have a clear stepwise ordering, namely the passage from any number n to its immediate successor $n+1$. This order exhausts the domain, in the sense that every positive integer may be obtained by applying the step finitely many times from the starting point.

Not all number systems are of this kind. The set $\mathbf{Z}$ of all integers, negative as well as positive, has no starting point under its natural ordering. The set $\mathbf{Q}$ of rational numbers not only lacks a starting point, but it also has the wrong kind of order. Conversely, the use of recursion and induction need not be confined to number systems. They can be carried out in any structure satisfying certain abstract conditions that make precise the vague requirement given above. But we will come to these later in the chapter.

4.2 Proof by Simple Induction on the Positive Integers

4.2.1 An Example

Suppose that we want to find a formula that identifies explicitly the sum of the first n positive integers. We might calculate a few cases, seeing that $1 = 1$, $1 + 2 = 3$, $1 + 2 + 3 = 6$, $1 + 2 + 3 + 4 = 10$, etc. In other words, writing $\Sigma\{i : 1 \le i \le n\}$ or more briefly just $f(n)$ for the sum of the first n integers, we see by calculation that $f(1) = 1$, $f(2) = 3$, $f(3) = 6$, $f(4) = 10$, etc. After some experimenting, we may hazard the conjecture (or read somewhere) that quite generally $f(n) = n \cdot (n{+}1)/2$. But how can we *prove* this?

If we continue calculating the sum for specific values of n without ever finding a counterexample to the conjecture, we may become more and more convinced that it is correct; but that will never give us a *proof* that it is so. For no matter how many specific instances we calculate, there will always be infinitely many still to come. We need another method – and that is supplied by simple induction. Two steps are needed.

- The first step is to note that the conjecture $f(n) = n \cdot (n{+}1)/2$ holds for the initial case that $n = 1$, i.e. that $f(1) = 1 \cdot (1{+}1)/2$. This is easy to verify, indeed immediate, since we have already noticed that $f(1) = 1$ and clearly also $1 \cdot (1{+}1)/2 = 1$. This step is known as the *basis* of the induction.

- The second step is prove a general statement: whenever the conjecture $f(n) = n \cdot (n{+}1)$ holds for $n = k$, then it holds for $n = k{+}1$. In other words, we show that for all positive integers k, if $f(k) = k \cdot (k{+}1)/2$ then $f(k{+}1) = (k{+}1) \cdot (k{+}2)/2$. This *general if-then* statement is known as the *induction step* of the proof. Notice how the equality in its consequent is formulated by substituting $k{+}1$ for k in the equality in its antecedent.

Taken together, these two are enough to establish our original conjecture. The first step shows that the conjecture holds for the number 1. The induction step may then be applied to that to conclude that it also holds for 2; but it may also be applied to *that* to conclude that the conjecture also holds for 3, and so on for any positive integer n. We don't actually have to perform all these applications one by one – indeed, we couldn't do so, for there are infinitely many of them. But we have a guarantee, from the induction step, that each of these applications could be made.

In the example, how do we go about proving the induction step? As it is a universally quantified conditional statement about all positive integers k, we *let k* be an arbitrary positive integer, *suppose* the antecedent to be true, and *show* that the consequent must also be true. In detail:

Let k be an arbitrary positive integer. *Suppose* $f(k) = k \cdot (k+1)/2$. We need to *show* that $f(k+1) = (k+1) \cdot (k+2)/2$. Now:

$$f(k+1) = 1 + 2 + \ldots + k + (k+1) \qquad \text{by the definition of the function } f$$
$$= (1 + 2 + \ldots + k) + (k+1) \qquad \text{arranging brackets}$$
$$= f(k) + (k+1) \qquad \text{by the definition of the function } f \text{ again}$$
$$= k \cdot (k+1)/2 + (k+1) \qquad \textit{by the supposition } f(k) = k \cdot (k+1)/2$$
$$= [k \cdot (k+1) + 2(k+1)]/2 \qquad \text{by elementary arithmetic}$$
$$= (k^2 + 3k + 2)/2 \qquad \text{by elementary arithmetic}$$
$$= (k+1) \cdot (k+2)/2 \qquad \text{by elementary arithmetic}$$

which completes the proof of the induction step, and thus of the proof as a whole! The key link in the chain of equalities is the italicized one, where we apply the supposition.

4.2.2 The Principle Behind the Example

The rule used in this example is called the *simple principle of mathematical induction*, and may be stated as follows. Consider any property that is meaningful for positive integers. To prove that every positive integer has the property, it suffices to show:

Basis: The least positive integer 1 has the property,

Induction step: Whenever a positive integer k has the property, then so does $k + 1$.

The same principle may be stated in terms of sets rather than properties. Consider any set $A \subseteq \mathbf{Z}^+$. To establish that $A = \mathbf{Z}^+$, it suffices to show:

Basis: $1 \in A$,

Induction step: Whenever a positive integer $k \in A$, then also $k + 1 \in A$.

Usually, checking the basis is a matter of trivial calculation. Establishing the induction step is carried out in the manner of the example: we let *let k* be an arbitrary positive integer, *suppose* that k has the property (that $k \in A$), and *show* that $k+1$ has the property (that $k+1 \in A$). Of course, tougher problems require more sweat in the last part, but still within this general framework.

Important terminology: Within the induction step, the supposition that k has the property, is called the *induction hypothesis*. What we set out to show from that supposition, i.e. that $k+1$ has the property, is called the *induction goal*.

EXERCISE 4.2.1 (WITH SOLUTION)

Use the principle of induction over the positive integers to show that for every positive integer n, the sum of the first n odd integers is n^2.

Solution: Write $f(n)$ for the sum of the first n odd integers, i.e. for $1 + 3 + \ldots + (2\,n{-}1)$. We need to show that $f(n) = n^2$ for every positive integer n.

Basis: We need to show that $f(1) = 1^2$. But clearly $f(1) = 1 = 1^2$ and we are done.

Induction step: Let k be any positive integer, and suppose (induction hypothesis) that the property holds when $n = k$, i.e. suppose that $f(k) = k^2$. We need to show (induction goal) that it holds when $n = k{+}1$, i.e. that $f(k{+}1) = (k{+}1)^2$. Now:

$$
\begin{aligned}
f(k+1) &= 1 + 3 + \ldots + (2\,k - 1) + (2(k+1) - 1) \quad \text{by definition of } f \\
&= (1 + 3 + \ldots + (2\,k - 1)) + (2(k+1) - 1) \, \text{arranging brackets} \\
&= f(k) + 2(k+1) - 1 \qquad\qquad \text{also by definition of } f \\
&= k^2 + 2(k+1) - 1 \qquad\qquad \textit{by the induction hypothesis} \\
&= k^2 + 2\,k + 1 \qquad\qquad\quad\; \text{by elementary arithmetic} \\
&= (k+1)^2 \qquad\qquad\qquad\; \text{again by elementary arithmetic.}
\end{aligned}
$$

What if we wish to prove that every *natural* number has the property we are considering? We proceed in exactly the same way, except that we start with 0 instead of 1:

Basis: The least natural number 0 has the property,

Induction step: Whenever a natura number k has the property, then so does $k{+}1$.

In the language of sets:

Basis: $0 \in A$,

Induction step: Whenever a natural number $k \in A$, then also $k + 1 \in A$.

Sometimes it is more transparent to state the induction step in an equivalent way using subtraction by one. For natural numbers, in the language of properties:

Induction step : For every natural number $k > 0$, if $k{-}1$ has the property, then so does k.

In the language of sets:

Induction step: For every natural number $k > 0$, if $k-1 \in A$, then also $k \in A$.

Note carefully the proviso $k > 0$: this is needed to ensure that $k-1$ is a natural number when k is. If we are inducing over the positive integers only, then of course the proviso is $k > 1$.

EXERCISE 4.2.2 (WITH SAMPLE SOLUTION)

Take the last formulation of the induction step, and state it in (equivalent) contrapositive form.

Solution: For every natural number $k > 0$, if $k \notin A$, then also $k-1 \notin A$.

EXERCISE 4.2.3 (WITH SOLUTION)

In Chapter 3, we used the pigeonhole principle to show that in any reception attended by $n \geq 1$ people, if everybody shakes hands just once with everyone else, then there are $n \cdot (n-1)/2$ handshakes. Show this again using the simple principle of induction over the positive integers.

Solution:

Basis: We show this for the case $n = 1$. In this case there are 0 handshakes, and $1 \cdot (1-1)/2 = 0$, so we are done.

Induction step: Let k be any positive integer, and suppose (induction hypothesis) that the property holds when $n = k$, i.e., suppose that when there are just k people, there are just $k \cdot (k-1)/2$ handshakes. We need to show (induction goal) that it holds when $n = k+1$, in other words, that when there are just $k+1$ people, there are $(k+1) \cdot ((k+1)-1)/2 = k \cdot (k+1)/2$ handshakes.

Consider any one of these $k+1$ people, and call this person a. Then (by an exercise fom the end of Chapter 1) the total number of handshakes is equal to the number involving a plus the number not involving a. Clearly, since a shakes hands just once with everyone else, there are just k handshakes of the former kind. Since there are just k people in the reception other than a, we know by the induction hypothesis that there are $k \cdot (k-1)/2$ handshakes of the latter kind. Thus there is a total of $k + k \cdot (k-1)/2$ handshakes. It remains to check by elementary arithmetic that $k + k \cdot (k-1)/2 = k \cdot (k+1)/2$, as follows:

$$k + k \cdot (k-1)/2 = (2\,k + k \cdot (k-1))/2$$
$$= (2\,k + k^2 - k))/2$$
$$= (k^2 + k))/2$$
$$= k \cdot (k+1)/2 \text{ as desired.}$$

This exercise illustrates the fact that a problem can sometimes be tackled in two ways – either directly as in Chapter 3, or by induction as here. Sometimes one is quicker, sometimes the other.

It also illustrates the need to keep clearly in mind what the induction hypothesis and induction goal are. Unless they are made explicit, it is easy to become confused. There is rarely any difficulty with the basis of an induction, but students often get into a mess with the induction step because they have not identified clearly what it is, and what its two components (hypothesis, goal) are.

Finally, it is good practice to separate in one's mind the general strategy of the induction from whatever numerical calculations may come up within its execution.

Alice Box: n versus k

Alice: Why do you use the variable n when you state the general proposition to be proven by induction, but the variable k when you prove the induction step?

Hatter: Just convention. We could use any other two variables.

Alice: Why not just one, say n?

Hatter: With careful formulation it could be done. But it is easy to get mixed up if you do that. It is usually easier to write (and read) when two distinct, conventionally chosen, variables like n and k are used.

4.3 Definition by Simple Recursion on the Natural Numbers

We now dig below the material of the preceding section. Roughly speaking, one can say that *underneath every inductive proof lurks a recursive definition*. In particular, when f is a function on the natural numbers (or positive integers) and we can prove inductively something about its behaviour, then f itself may be defined (or at least characterized) in a recursive manner.

For an example, consider again the function f used in the first example of this chapter. Informally, $f(n)$ was understood to be the sum $1 + 2 + 3 + \ldots + n$ of the first n positive integers. The reason why induction could be applied to prove that $f(n) = n.(n+1)/2$ is that the function f can itself be defined inductively or, as usually said in this context, *recursively*.

What would such a definition look like? As one would expect, it consists of a *basis* giving the value of f for the least positive integer argument 1, and a *recursive (or inductive) step* giving the value of f for any argument $n+1$ in terms of its value for argument n. Thus, in a certain sense the function is being defined in terms of itself, but in a non-circular manner: the value $f(n+1)$ is defined in terms of $f(n)$ with the lesser argument n. Specifically, in our example, the basis and recursive step are as follows:

$$\textit{Basis of definition:} \quad f(1) = 1$$
$$\textit{Recursive step of definition:} \quad f(n+1) = f(n) + (n+1).$$

This way of expressing the recursive step, using addition by 1, is sometimes called recursion by *pattern-matching*. Another way of writing it uses subtraction by 1, in our example as follows:

$$\textit{Recursive step of definition}: \quad \text{when } n > 1 \text{ then } f(n) = f(n-1) + n.$$

Other ways of writing recursive definitions are also current among computer scientists. In particular, one can think of the basis and induction step as being *limiting* and *principle* cases respectively, writing in our example:

$$\text{If } n = 1 \text{ then} f(n) = 1$$
$$\text{If } n > 1 \text{ then} f(n) = f(n-1) + n.$$

This can also be expressed in the popular *if-then-else* form:

$$\text{If } n = 1 \text{ then} f(n) = 1$$
$$\text{Else } f(n) = f(n-1) + n$$

And some computer scientists like to abbreviate this further to the ungrammatical declaration:

$$f(n) = \text{if } n = 1 \text{ then } 1 \text{ else} f(n-1) + n$$

which can look like mumbo-jumbo to the uninitiated.

All of these formulations are equivalent. You will meet each of them in applications, and so should recognize them. The choice is partly a matter of personal preference, partly a question of which one allows you to get on with the problem in hand with the least clutter.

EXERCISE 4.3.1 (WITH PARTIAL SOLUTION)

(a) Define recursively the following functions $f(n)$ on the positive integers:

   (i) The sum $1 + 3 + \ldots + (2n-1)$ of the first n odd integers

   (ii) The sum $2 + 4 + \ldots + 2n$ of the first n even integers

(b) Show by induction that the sum of the first n even integers equals $n \cdot (n+1)$.

(c) Give another proof of (b) without a fresh induction, but making use of the already established fact that the sum of the first n positive integers is $n \cdot (n+1)/2$.

(d) Define recursively the function that takes a natural number n to 2^n. This is called the *exponential function*.

(e) Define recursively the product of the first n positive integers. This is known as the *factorial function*, written $n!$ and pronounced 'n factorial'.

(f) Use the recursive definition of the functions concerned to show that for all $n \geq 4$, $n! > 2^n$. *Hint*: Here the basis will concern the case that $n = 4$.

   Comments on (d–f) The factorial and exponentiation functions are both very important in computer science, as indications of the alarming way in which many processes can grow in size (length, memory requirements, etc) as inputs increase. In Chapter 1 we saw that exponentiation already gives unmanageable rates of growth; the exercise shows that factorial is worse.

Solution to (d–f):

(d) Basis of definition: $2^0 = 1$. Recursive step of definition: $2^{n+1} = 2 \cdot 2^n$.

(e) Basis of definition: $1! = 1$. Recursive step of definition: $(n+1)! = (n+1) \cdot n!$

(f) Basis of proof: We need to show that $4! > 2^4$. This is immediate by calculation, with $4! = 1 \cdot 2 \cdot 3 \cdot 4 = 24 > 2^4 = 2 \cdot 2 \cdot 2 \cdot 2 = 16$. Induction step of proof: Let k be any positive integer with $k \geq 4$. Suppose that $k! > 2^k$ (induction hypothesis). We need to show that $(k+1)! > 2^{k+1}$ (induction goal). This can be done as follows:

   $(k + 1)! = (k + 1) \cdot k!$ by definition of the factorial function

   $> (k + 1) \cdot 2^k$ by the induction hypothesis

   $> 2 \cdot 2^k$ since $k \geq 4$

   $= 2^{k+1}$ by definition of the exponential function.

4.4 Evaluating Functions Defined by Recursion

Go back yet again to the first function that we defined by recursion, the sum $f(n)$ of the first n positive integers. Suppose that we want to calculate $f(7)$. There are basically two ways of doing it.

One very obvious way is *bottom-up*. We first calculate $f(1)$, use it to get $f(2)$, use it for $f(3)$, and so on. Thus we have:

$$f(1) = 1$$
$$f(2) = f(1) + 2 = 1 + 2 = 3$$
$$f(3) = f(2) + 3 = 3 + 3 = 6$$
$$f(4) = f(3) + 4 = 6 + 4 = 10$$
$$f(5) = f(4) + 5 = 10 + 5 = 15$$
$$f(6) = f(5) + 6 = 15 + 6 = 21$$
$$f(7) = f(6) + 7 = 21 + 7 = 28.$$

There is one application of the base clause and six of the recursion clause. Each application is accompanied, as needed, by arithmetic simplification. Each of the seven steps fully eliminates the function sign f and provides us with a specific numeral.

The other way of proceeding, at first a little less obvious, is *top-down*, also known as the process of *unfolding* the recursive definition or *tracing* the function. It is as follows:

$$
\begin{aligned}
f(7) &= f(6) + 7 \\
&= (f(5) + 6) + 7 \\
&= ((f(4) + 5) + 6) + 7 \\
&= (((f(3) + 4) + 5) + 6) + 7 \\
&= ((((f(2) + 3) + 4) + 5) + 6) + 7 \\
&= (((((f(1) + 2) + 3) + 4) + 5) + 6) + 7 \\
&= (((((1 + 2) + 3) + 4) + 5) + 6) + 7 \\
&= ((((3 + 3) + 4) + 5) + 6) + 7 \\
&= (((6 + 4) + 5) + 6) + 7 \\
&= ((10 + 5) + 6) + 7 \\
&= (15 + 6) + 7 \\
&= 21 + 7 \\
&= 28
\end{aligned}
$$

In this calculation, we begin by writing the expression $f(7)$, make a substitution for it as authorized by the recursion clause of the definition, then in that expression substitute for $f(6)$ etc, until after six steps we can at last apply the basis to $f(1)$ to emerge with a fully numerical expression not containing f. At that point we begin simplifying the numerical expression until, in another six steps, we have got it in the form of a standard numeral.

Which is the better way to calculate? In this example, it does not make a significant difference. Indeed, quite generally, when the function is defined using simple recursion of the kind described in this section, the two modes of calculation will be of essentially the same length. The second one looks longer, but that is because we have left all arithmetic simplifications to the end (as is customary when working top-down) rather than doing them as we go. Humans often prefer the first mode, for the psychological reason that it gives us something 'concrete' at each step; but a computer would not care.

Nevertheless, when the function is defined by more sophisticated forms of recursion, which we will describe in the following sections, the situation is very different. It can turn out that one, or the other, of the two modes of evaluation is dramatically more economical in resources of memory or time.

Such economies are of little interest to the traditional mathematician, but they are of great importance for the computer scientist. They may make the difference between a feasible calculation procedure and one that, in a given state of technology, is quite unfeasible.

EXERCISE 4.4.1

Evaluate 6! bottom-up and then again top-down (unfolding).

4.5 Cumulative Induction and Recursion

We now turn to some rather more sophisticated forms of recursive definition and inductive proof.

4.5.1 Recursive Definitions Reaching Back More Than One Unit

In definitions by simple recursion such as that of the factorial function, we reached back only one notch at a time. In other words, the recursion step defined $f(n)$ out of $f(n-1)$. But sometimes we want to reach back further. A famous example is the

Fibonacci function on the natural numbers, so named after an Italian mathematician who considered it in the year 1202, and which has found surprisingly many applications in computer science, biology and elsewhere. To define it we use recursion. The basis has two components:

$$F(0) = 0$$
$$F(1) = 1.$$

So far just like the identity function. But then, in the recursion step Fibonacci takes off:

$$F(n) = F(n-1) + F(n-2) \text{ whenever } n \geq 2.$$

The Fibonacci function illustrates the way in which a top-down evaluation by unfolding can lead to great inefficiencies in computation. Beginning the computation of $F(8)$ top-down we have to make the following calls:

Table 4.1 Calls when computing F(8) top-down.

F(8)															
F(7)								F(6)							
F(6)				F(5)				F(5)				F(4)			
F(5)		F(4)		F(4)		F(3)		F(4)		F(3)		F(3)		F(2)	
$F(4)$	$F(3)$	$F(3)$	$F(2)$	$F(3)$	$F(2)$	$F(2)$	$F(1)$	$F(3)$	$F(2)$	$F(2)$	$F(1)$	$F(2)$	$F(1)$	$F(1)$	$F(0)$
etc															

In this table, each cell in a row is split into two in the row below, following the recursive rule for the Fibonacci function, so that value of each cell equals the sum of the values in the two cells immediately below. The table is not yet complete – it hasn't even reached the point where all the letters F are eliminated and the arithmetic simplifications begin – but it is already clear that there is a great deal of repetition. The value of $F(6)$ is calculated twice, that of $F(5)$ three times, $F(4)$ five times, $F(3)$ eight times (including the one still to come in the next row). Indeed, when calculating $F(n)$, the number of times each $F(k)$ is calculated itself follows a Fibonacci function run backwards from $n+1$. For example, to calculate $F(8)$, each $F(k)$ for $k \leq 8+1 = 9$ is calculated $F(9-k)$ times. Unless partial calculations are saved and recycled in some manner, the inefficiency is high.

EXERCISE 4.5.1

(a) Express the definition of the Fibonacci function in pattern-matching form, and then again in *if-then-else* form.

(b) Carry out a bottom-up evaluation of $F(8)$.

On the other hand, there are cases in which evaluation bottom-up can be less efficient. Here is a simple example. Define f as follows:

Basis: $f(0) = 2$

Recursion step: $f(n) = f(n-1)^n$ for odd $n > 0$, $f(n) = f(n/2)$ for even $n > 0$

To calculate $f(8)$ by unfolding is quick and easy: $f(8) = f(4) = f(2) = f(1)$ by three applications of the second case of the recursion step and finally $f(1) = f(0)^1 = 2^1 = 2$ by the first clause of the recursion step and the basis. On the other hand, if we were to calculate $f(8)$ bottom-up without introducing shortcuts, we would pass through each of $f(0)$ to $f(8)$ doing unnecessary work on the odd values. We will see a more dramatic example shortly.

4.5.2 Proof by Cumulative Induction

There is no limit to how far a recursive definition may reach back when defining $f(n)$, and so it is useful to have a form of proof that permits us to do the same. It is called *cumulative induction*, sometimes also known as *course-of-values* induction.

Formulated for the natural numbers, in terms of properties, the *principle of cumulative induction* is as follows: To show that every natural number has a certain property, it suffices to show the basis and induction step:

Basis: 0 has the property.

Induction step: For every natural number k, if every natural number $j < k$ has the property, then k itself also has the property.

The same idea can be used when our induction begins higher than 0. To show that every natural number $n \geq a$ has a certain property, it suffices to show the corresponding basis and induction step:

Basis: a has the property.

Induction step: For every natural number $k \geq a$, if every natural number j with $a \leq j < k$ has the property, then k itself also has the property.

EXERCISE 4.5.2 (WITH SOLUTION)

(a) Formulate the induction step of the principle of cumulative induction contrapositively.

(b) Use cumulative induction to show that every positive integer $n \geq 2$ is the product of (one or more) prime numbers.

Solution:

(a) For every natural number k, if k lacks the property, then there is a natural number $j < k$ that lacks the property.

(b) *Basis*: We need to show that 2 has the property. Since 2 is itself prime, we are done.

Induction step: Let k be any positive integer with $k \geq 2$. Suppose (induction hypothesis) that every positive integer j with $2 \leq j < k$ has the property. We need to show (induction goal) that k also has it. There are two cases to consider. If k is prime, then we are done. On the other hand, if k is not prime, then by the definition of prime numbers $k = a \cdot b$ where a, b are positive integers ≥ 2. Hence $a < k$ and $b < k$. So we may apply the induction hypothesis to get that each of a, b has the property, i.e. is the product of (one or more) prime numbers. Hence their product is too, and we are done.

In this exercise we have kept things brief by speaking of 'the property' when we mean 'the property of being the product of two prime numbers'. The example was simple enough for it to be immediately clear what property we are interested in; in more complex examples you are advised to take the precaution of stating the property in full at the beginning of the proof.

Logic Box: Vacuous basis for cumulative induction

An interesting feature of proof by cumulative induction is that, strictly speaking, the basis is also covered by the induction step, and so is redundant! This contrasts with simple induction, where the basis is quite independent of the induction step and always needs to be established separately.

The reason for this redundancy is that there are no natural numbers less than 0. Hence, vacuously, every natural number $j < 0$ has whatever property is under consideration, so that by the induction step, 0 itself has the property.

Nevertheless, even for cumulative induction it is customary to formulate the basis explicitly and to check it out separately. Even though the basis is

(Continued)

> *Alice Box:* (Continued)
>
> redundant, this is good practice. It acts as a double-check that we did not overlook some peculiarity of zero when establishing the induction step.

EXERCISE 4.5.3

Use cumulative induction to show that for every positive integer n, $F(n)$ is even iff n is divisible by 3, where F is the Fibonacci function.

4.5.3 Simultaneous Recursion and Induction

When a function has more than one argument, its definition will have to take account of both of them. If the definition is recursive, sometimes we can get away with recursion on just one of the arguments, holding the others as parameters. A simple example is the recursive definition of multiplication over the natural numbers, using addition, which can be expressed as follows:

$$Basis\colon \ (m \cdot 0) = 0$$

$$Recursion \ step\colon \ m \cdot (n + 1) = (m \cdot n) + m$$

The equality in the recursion step is usually taught in school as a *fact* about multiplication, which is assumed to have been defined or understood in some other way. In Peano's axiomatization of arithmetic in the context of first-order logic, the equality is treated as an *axiom* of the system. But here we are seeing it as the recursive part of a *definition* of multiplication, given addition. The same mathematical edifice can be built in many ways!

EXERCISE 4.5.4

Give a recursive definition of the power function that takes a pair (m, n) to m^n, using recursion on the second argument only.

When a two-argument function is defined in this way, then inductive proofs about it will tend to follow the same pattern, with induction carried out on the argument that was subject to recursion in the definition.

But sometimes we need to define functions of two (or more) arguments with recursions on each of them. This is called *definition by simultaneous recursion*. A famous example is the *Ackermann function*. It has two arguments and, in one of its variants (due to Rósza Péter), is defined as follows:

$$A(0, n) = n + 1$$
$$A(m, 0) = A(m - 1, 1) \text{ for } m > 0$$
$$A(m, n) = A(m - 1, A(m, n - 1)) \text{ for } m, n > 0$$

Alice Box: Basis and induction step

Alice: One minute! What is the basis and what is the recursion step here? There are three clauses.

Hatter: The first clause gives the basis. Although it covers infinitely many cases, and uses a function on the right hand side, the function A that is being defined does not appear on the right. The other two clauses taken together make up the recursion step. One covers the case $m > 0$ while $n = 0$; the other deals with the case when both $m, n > 0$. The distinguishing feature that marks them as parts of the recursion step is the appearance of A on the right.

The reason why the Ackermann function is famous is its spectacular rate of growth. For $m < 4$ it remains leisurely, but when $m \geq 4$ it accelerates dramatically, much more than either the exponential or the factorial function. Even $A(4,2)$ is about $2 \cdot 10^{19728}$. This gives it a theoretical interest: although the function is computable, it grows faster than any function in the class of so-called 'primitive recursive' functions, which for a short time were thought to exhaust the computable functions.

But what interests us about the function here, is the way in which the last clause of the definition makes the value of $A(m,n)$ depend on the value of the same function for the first argument diminished by 1, but paired with a value of the second argument *that might be larger than* n. As the function picks up speed, to calculate the value of $A(m,n)$ for a given m may require prior calculation of $A(m{-}1,n')$ for an extremely large $n' > n$.

Indeed, given the way in which the recursion condition reaches 'upwards' on the second variable, it is not immediately obvious that the clauses taken together really succeed in defining a unique function. It can, however, be shown that they do, by introducing a suitable well-founded ordering on $\mathbf{N}^2$, and using the principle of well-founded recursion. We will introduce those concepts in a later section of this chapter.

This function also illustrates the differences that can arise between calculating bottom-up (alias forwards) or top-down (alias backwards, or by unfolding), in the sense described in the preceding section. The best way to appreciate the difference is to do the following exercise.

EXERCISE 4.5.5

Calculate the value of $A(2,3)$ bottom-up. Then calculate it top-down (unfolding).

This time the bottom-up approach comes off worse. The difficulty is that, as already remarked, to calculate $A(m,n)$ for a given m may require prior calculation of $A(m{-}1,n')$ for a certain $n' > n$. We don't know in advance how large this n' will be, and it is very inefficient to plough through all the values of n' until we reach the one needed.

4.6 Structural Recursion and Induction

We now come to the form of recursion and induction that is perhaps the most frequently used by computer scientists and logicians – *structural*. It can be justified or replaced by recursion and induction on the natural numbers, but may conveniently be used without ever mentioning them. If this sounds like magic, read on.

We introduce structural recursion/induction in three stages, remembering the rough dictum that behind every inductive proof lurks a recursive definition. First, we look at the business of *defining sets* by structural recursion. That will not be difficult, because in fact we have already been doing this, without great attention to the recursive aspect, back in Chapters 2 and 3. Then we will turn to the task of *proving things about these sets* by structural induction, which will also be quite straightforward. Finally, we come to the rather tricky part, examining the task of taking a recursively defined set and *defining a function with it as domain*, by structural recursion. This is where care has to be taken; such definitions are legitimate only when a special constraint is satisfied.

4.6.1 Defining Sets by Structural Recursion

In Chapters 2 and 3 we introduced the notions of the image and the closure of a set under a relation or function. We now make intensive use of them. We begin by recalling their definitions, generalizing a little from binary (i.e. two-place) relations to relations of any finite number $n \geq 2$ of places.

Let A be any set, and let R be any relation (of at least two places) over the local universe within which we are working. Since m-argument functions are $(m{+}1)$-place relations, this covers functions of one or more arguments as well.

The *image* of A under the $(m+1)$-place relation R is defined by putting $x \in R(A)$ iff there are $a_1, \ldots, a_m \in A$ with $(a_1, \ldots a_m, x) \in R$. In the case that R is an m-argument function f, this is equivalent to saying: $x \in f(A)$ iff there are $a_1, \ldots, a_m \in A$ with $x = f(a_1, \ldots a_m)$.

The concept of image is not yet recursive; that comes with the closure $R[A]$, which for brevity we write as A^+. We saw that it can be defined in either of two ways.

The *way of union* (bottom up) is by making a cumulative recursion on the natural numbers, then take the union of all the sets thus produced:

$$A_0 = A$$
$$A_{n+1} = A_n \cup R(A_n), \text{for each natural number } n$$
$$A^+ = \cup\{A_n : n \in \mathbf{N}\}.$$

The *way of intersection* ('top down') dispenses with numbers altogether. We say that a set X is *closed* under R iff $R(X) \subseteq X$. The *closure* A^+ of A under the relation R is defined to be the *intersection of all sets* $X \supseteq A$ that are closed under R. That is, $A^+ = \cap\{X : A \subseteq X \supseteq R(X)\}$.

So defined, A^+ is in fact *the least* superset of A that is closed under all the relation R. On the one hand, being the intersection of all the X, it is clearly included in each X; on the other hand, it is easy to check that it is itself also closed under the relation R.

EXERCISE 4.6.1

(a) Verify the last point made, that the intersection of all sets $X \supseteq A$ that are closed under R is itself closed under R.

(b) For this exercise, to avoid confusion, use $A^\cup$ as temporary notation for A^+ defined bottom-up, and $A^\cap$ as temporary notation for A^+ defined top-down. Show that $A^\cup = A^\cap$ by establishing the two inclusions separately, as follows:

  (i) Use cumulative induction on the natural numbers to show that $A^\cup \subseteq A^\cap$.

  (ii) Show that $A^\cup$ is closed under R and use this to get $A^\cap \subseteq A^\cup$.

The definition can evidently be extended to cover an arbitrary collection of relations, rather than just one. The *closure* A^+ of A under a collection $\{R_i\}_{i \in I}$ of relations is defined to be the *intersection of all sets* $X \supseteq A$ that are closed under all the relations in the collection.

A set is said to be *defined by structural recursion* whenever it is introduced as the closure of some set (referred to as the *basis*, or initial set) under some collection of relations (often called *constructors* or *generators*).

In this context, it is intuitively helpful to think of each $(m+1)$-place relation R as a *rule*: if $(a_1, \ldots a_m, x) \in R$ then, when building A^+, we are authorized to pass from items $a_1, \ldots a_m$ already in A^+, to put x in A^+ too. Thus, A^+ may be described as *the closure of A under a collection of rules*. This way of speaking often makes formulations more vivid.

We illustrate the concept by several examples drawn from computer science and logic.

- The *notion of a string*, already mentioned informally in an Alice box of Chapter 3. It is of central importance for formulating e.g. the theory of finite state machines. Let A be any alphabet, consisting of elementary signs. Let λ be an abstract object, distinct from all the letters in A, which is understood to represent the empty string. The set of all strings over A, conventionally written as A^*, is the *closure* of $A \cup \{\lambda\}$ under the rule of concatenation, i.e. the operation of taking two strings s, t and forming their concatenation by writing s immediately followed by t.

- We can define *specific sets of strings* by structural recursion. For instance, a string over an alphabet A is said to be a *palindrome* iff it reads the same from each end. Can we give this informal notion a precise recursive definition? Very easily! The empty string λ reads the same way from each end, and is the shortest even palindrome. Each individual letter in A reads the same way from left and from right, and so these are the shortest odd palindromes. All other palindomes may be obtained by successive symmetric flanking. So we may take the set P of palindromes to be the *closure* of the set $\{\lambda\} \cup A$ under the rule permitting passage from a string s to a string xsx for any $x \in A$. In other words, it is the least set S including $\{\lambda\} \cup A$ such that $xsx \in S$ for any $s \in S$ and $x \in A$.

- Logicians also work with symbols, and constantly define sets by structural recursion. In particular, the set of *formulae of classical propositional logic* (or any other logical system) is defined as the *closure* of an initial set A under some operations. In this case, A is a set of proposition letters. It is closed under the rules for forming compound formulae by means of the logical connectives allowed, e.g. $\neg, \wedge, \vee$ (with parentheses to ensure unambiguous reading). So defined, the set of propositional formulae is a particular subset of the set B^* of all strings in the alphabet $B = A \cup \{\neg, \wedge, \vee\} \cup \{(,)\}$. In case you have trouble reading it, $\{(,)\}$ is the set consisting of the left and right parentheses.

- The set of *theorems in a formal logical system* can also be defined by structural recursion. It is the *closure A^+* of some initial set A of formulae (known as the *axioms* of the system) under certain functions or relations between formulae (known as *derivation rules* of the system).

- Algebraists also use this kind of definition, even though they are not, in general dealing with strings of symbols. For example, if A is a subset of an algebra, then the *subalgebra generated by A* has as its carrier (i.e. underlying set) the *closure A^+* of A under the operations of the algebra.

In all these cases, it is perfectly possible to use natural numbers as indices for successive sets A_0, A_1, $A_2,\ldots$ in the generation of the closure, and replace the structural definition by one that makes a cumulative recursion on the indices. Indeed, this is a fairly common style of presentation, and in some contexts has its advantages. But in general, the use of indices and appeal to induction on numbers is an unnecessary detour.

Alice Box: Defining the set **N** *recursively*

Alice: My friend studying the philosophy of mathematics tells me that even the set of natural numbers may be defined by structural recursion. This, he says, is the justification for induction over the integers. Is that possible?

Hatter: We can define the natural numbers in that way, if we are willing to identify them with sets. There are many ways of doing it. For example, we can take **N** to be the least set X that contains $\emptyset$ and is closed under the operation taking each set S to $S \cup \{S\}$. In this way arithmetic is reduced to set theory.

Alice: Does that *justify* induction over the natural numbers?

Hatter: When arithmetic is axiomatized in its own terms, without reducing it to anything else, induction is simply treated as an axiom. In that context, it is not justified formally at all. When arithmetic is reduced to set theory, say along the lines that I just mentioned, induction need no longer be treated as an axiom – it can be proven. Which the best way of doing things depends on your philosophy of mathematics. But discussing these matters would take us too far off our track.

EXERCISE 4.6.2 (WITH SOLUTION)

(a) Define by structural recursion the set of even palindromes over an alphabet A, i.e. the palindromes with an even number of letters.

(b) Let φ be any formula of propositional logic. Define recursively the set of all formulae of which φ is a *subformula*. *Hint*: Assume that the set of all formulae is itself defined recursively using the connectives $\neg, \wedge, \vee$, and abbreviate formulae by omitting brackets as convenient.

Solution:

(a) The set of even palindromes over an alphabet A is the closure of $\{\lambda\}$ under the same rule, i.e. passage from a string s to a string xsx for any $x \in A$.

(b) Consider the rule allowing passage from any propositional formula α to $\neg\alpha$, and the rule allowing passage from α to any one of $\alpha\wedge\beta$, $\beta\wedge\alpha$, $\alpha\vee\beta$ or $\beta\vee\alpha$ for any formula β whatsoever. Then the desired set is the closure A^+ of $A = \{\varphi\}$ under these two rules.

4.6.2 Proof by Structural Induction

We have seen that the procedure of defining a set by structural recursion, i.e. as the closure of a set under given relations or functions, is pervasive in computer science, logic and abstract algebra. Piggy-backing on this mode of definition is a mode of demonstration that we will now examine – proof by structural induction.

Let A be a set, of any items whatsoever, let A^+ be the closure of A under a collection $\{R_i\}_{i\in I}$ of relations. Consider any property that we would like to show holds of all elements of A^+. We say that a relation R *preserves the property* iff whenever $a_1, \ldots a_m$ have the property and $(a_1, \ldots a_m, x) \in R$, then x also has it. When R is a function f, this amounts to requiring that whenever $a_1, \ldots a_m$ has the property then $f(a_1, \ldots a_m)$ also has the property.

The *principle of proof by structural induction* may now be stated as follows. Again, let A be a set, and A^+ the closure of A under a collection $\{R_i\}_{i\in I}$ of relations. To show that every element of A^+ has a certain property, it suffices to show:

Basis: Every element of A has the property.

Induction step: Each relation $R \in \{R_i\}_{i\in I}$ preserves the property.

Justification of the principle is almost immediate given the definition of the closure A^+. Let X be the set of all items that have the property in question. Suppose that both basis and induction step hold. Since the basis holds, $A \subseteq X$. Since the induction step holds, X is closed under the relations R_i. Hence by the definition of A^+ as the *least* set with those two features, we have $A^+ \subseteq X$, i.e. every element of A^+ has the property, as desired.

For an example of the application of this principle, suppose we want to show that every even palindrome has the property that every letter that occurs in it occurs an even number of times. Recall that the set of even palindromes was defined as the closure of $\{\lambda\}$ under passage from a string s to a string xsx where $x \in A$. So we need only show two things: that λ has the property in question, and that whenever s has it then xsx does too. The former holds vacuously, since there are no letters in λ (remember, λ is not itself a letter). The latter is trivial, since the passage from s to xsx adds two more occurrences of the letter x without disturbing the other letters.

EXERCISE 4.6.3

(a) Use proof by structural induction to show that in any (unabbreviated) formula of propositional logic, the number of left brackets equals the number of right brackets.

(b) Let A be any set of formulae of propositional logic, and let R be the three-place relation of *detachment* (alias *modus ponens*) defined by putting $(a,b,c) \in R$ iff b is the formula $a \rightarrow c$. Use structural induction to show that every formula in A^+ is a subformula of some formula in A.

4.6.3 Defining Functions by Structural Recursion on Their Domains

There is a difference between the two examples in the last exercise. In the first one, we begin with a set A of elementary letters, and the closing functions (forming negations, conjunctions, disjunctions) produce longer and longer formulae, and so always give us something fresh. But in the second example, the closing relation may sometimes give us a formula that is already in the initial set A, or already available at an earlier stage of the closing process: *it is not guaranteed always to give us something fresh*.

This difference is of no significance for structural induction as a method of proof, but it is very important, indeed vital, if we want to use structural recursion to define a function whose domain is a set already defined by structural recursion – a situation that arises quite frequently.

Suppose that we want to give a recursive definition of a function f whose domain is the closure A^+ of a set A under an injective function g. For example, we might want to define a function $f\colon A^+ \rightarrow \mathbf{N}$, as some kind of measure of complexity of the elements of the domain, by putting $f(a) = 0$ for all $a \in A$, and $f(g(a)) = f(a)+1$. Since g is injective, the second part of the definition makes sense, but is the definition as a whole legitimate?

Unfortunately, not always. The elements $g(a)$ introduced by the function g may not be fresh. It may turn out that $g(a)$ is already in the set A, so that $f(g(a))$ was already defined as 0 by the first part of the definition, with the result that it is defined twice in different ways.

To bring this out, we make our example completely concrete. Let $\mathbf{N}$ be the set of natural numbers and $g: \mathbf{N} \to \mathbf{N}$ the function of adding one except that $g(99)$ is 0. That is, $g(n) = n+1$ for $n \neq 99$, while $g(99) = 0$. Clearly, this function is injective. Put $A = \{0\}$. Then the closure A^+ of A under g is the set $\{0,\ldots,99\}$. Now suppose that we wish to define a function $f: A^+ \to \mathbf{N}$ by structural induction, putting $f(0) = 0$ and $f(g(n)) = f(n)+1$ for all $n \in A^+$. This definition is fine for values of $n < 99$, indeed it is the identity function for those values. But it breaks down at $n = 99$. The recursion step tells us that $f(g(99)) = f(99)+1 = 100$, but since $g(99) = 0$, the basis also forces us to say $f(g(99)) = f(0) = 0$. Although the basis and the recursion step make sense separately, they conflict, and we have not succeeded in defining a function!

Inductive proof and recursive definition, which went hand in hand up to now, thus come apart a little at this stage:

- Structural recursion, *as a method of defining a set* as the closure of an initial set under relations or functions, is always legitimate.

- Structural induction, *as a method of proof that all elements of a recursively defined set have a certain* property, is always a legitimate strategy.

- But structural recursion, *as a method of defining a function with a recursively defined set as its domain*, can fail to yield a function as desired.

This prompts the question: Is there a condition whose satisfaction guarantees that the definition succeeds, i.e. that the function is well-defined?

4.6.4 Condition for Defining a Function by Structural Recursion

Fortunately, analysis of examples like the above suggests a condition whose satisfaction eliminates the danger of failure. Fortunately too, the condition holds in many situations of concern to the computer scientist and logician. To simplify notation a little, we will also focus on the case that the closure is generated by functions.

Let A be a set and A^+ its closure under a collection $\{g_i\}_{i \in I}$ of functions. An element x of A^+ is said to be *uniquely decomposable* iff either (1) $x \in A$, and is not in the range of any of the functions g_i, or (2) $x \notin A$, and $x = g_i(y_1,\ldots,y_k)$ for a unique function g_i in the collection and a unique tuple $y_1,\ldots,y_k \in A^+$. Roughly speaking, this guarantees that there is a unique way in which x can have got into A^+. Because many instances of structural recursion in computer science or logic concern sets of expressions, this property is also often called *unique readability*.

Unique decomposability suffices to guarantee that a structural recursion on A^+ succeeds, i.e. that it defines a unique function with A^+ as domain. To be precise:

Principle of structural recursive definition: let A be a set, $\{g_i\}_{i \in I}$ a collection of functions, and A^+ the closure of A under the functions. Suppose that every element of A^+ is uniquely decomposable. Let X be any set, and let $f\colon A{\to}X$ be a given function on A into X. Then, for every collection $\{h_i\}_{i \in I}$ of functions h_i on appropriate powers of X into X, there is a unique function $f^+\colon A^+{\to}X$ satisfying the following recursively formulated conditions:

Case	Definition
Basis: $x \in A$	$f^+(x) = f(x)$
Recursion step: $x = g_i(y_1,\ldots,y_k)$ is the unique decomposition of x	$f^+(x) = h_i(f^+(y_1),\ldots,f^+(y_k))$

Another way of putting this principle, which will ring bells for readers who have done some abstract algebra, is as follows: We may legitimately extend a function $f\colon A{\to}X$ homomorphically to a function $f^+\colon A^+{\to}X$ if every element of A^+ is uniquely decomposable.

This is highly abstract, and may at first be rather difficult to follow. It may help to visualize it through the following diagram for the case that A is a singleton $\{a\}$, A^+ is its closure under just one function g with just one argument, so that there is also just one function h under consideration likewise with just one argument.

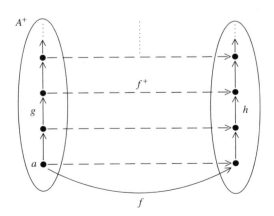

Figure 4.1 *Recursive structural definition.*

The good news is that when we look at specific applications of this principle, they are usually very natural – so much so that we sometimes use them without realizing it, taking the existence of unique extensions for granted.

For a very simple example from computer science, suppose we are given an alphabet A, and let A^+ be the set of all (finite) lists that can be formed from this alphabet by the operation *cons*, of prefixing a letter of the alphabet to an arbitrary list (see the Alice box 'n-tuples, sequences, strings and lists' in Section 3.5). We might define the *length* of a list recursively as follows, where $<>$ is the empty list:

Basis: $length(<>) = 0$

Induction step: If $length(l) = n$ and $a \in A$, then $length(al) = n + 1$.

The principle tells us that this definition is legitimate, because *each non-empty list has a unique decomposition into a head and a body*.

Essentially the same idea is involved when we define the logical *depth* of a formula of propositional logic. Suppose our formulae are built up using just negation, conjunction and disjunction. Intuitively, we say:

Basis: $depth(p) = 0$ for any elementary letter p

Induction step:

      Case 1. If $depth(\alpha) = m$ then $depth(\neg\alpha) = m + 1$

      Case 2. If $depth(\alpha) = m$ and $depth(\beta) = n$ then $depth(\alpha \wedge \beta)$

      $= depth(\alpha \vee \beta) = max(m, n) + 1$

and take it for granted that the function is well defined. The principle of structural recursive definition tells us that indeed it is so, because the formulae of propositional logic have unique decompositions under the operations of forming negations, conjunctions, and disjunctions (with suitable bracketing).

EXERCISE 4.6.4

(a) Let A be any alphabet, and A^* the set of all strings over A. Let $a \in A$ and $s \in A^*$. Intuitively, the substitution function $\sigma_{a,s}$ substitutes the string s for the letter a in all strings. Define this function by structural recursion. *Hint*: In the basis you will need to distinguish cases.

(b) In propositional logic, two formulae α, β are said to be tautologically equivalent, and we write $\alpha =||= \beta$, iff they receive the same truth-value as each other, under every possible assignment of truth-values to elementary letters. Three well-known tautological equivalences are double negation $\alpha =||= \neg\neg\alpha$ and the de Morgan principles $\neg(\alpha\wedge\beta) =||= \neg\alpha\vee\neg\beta$ and $\neg(\alpha\vee\beta) =||= \neg\alpha\wedge\neg\beta$. With these in mind, define by structural recursion a function that takes every propositional formula (built using the connectives $\neg,\vee,\wedge$) to a tautologically equivalent one in which negation is applied only to elementary letters. *Hint*: The recursion step will need to distinguish cases.

4.6.5 When the Unique Decomposition Condition Fails?

Can anything be done when the unique decomposition condition fails? Can we still define functions on the closure of A? Without unique decomposition, we can no longer do so in the way described in the principle of structural recursion. But we can still define a function to measure the depth of an element of A^+.

Recall from Section 3.2 that the closure of A under given relations may also be defined as the union $\cup\{A_n : n \in \mathbf{N}\}$ of a sequence of sets. Hence for each element a of the closure there is an $n \in \mathbf{N}$ with $a \in A_n$. For a given a of the closure there may be many such n, but since the set of all natural numbers is well-ordered, we know that there must be a *least* such n. We can therefore always define a function *depth*: $A^+ \rightarrow \mathbf{N}$ by putting $depth(a)$ to be the least n with $a \in A_n$. This definition will succeed, even when the operations constructing A^+ out of A do not satisfy the unique decomposition condition.

4.7 Recursion and Induction on Well-Founded Sets

The concept of a well-founded set provides the most general context for recursion and induction. It permits us to apply these procedures to any domain whatsoever – provided we have available a well-founded relation over the domain. Every other form can in principle be derived from it. In this section we give a glimpse of the bare essentials that need to be understood in order to be able to use it.

4.7.1 Well-Founded Sets

We begin by defining the notion of a well-founded relation over a set. Let W be any set, and $<$ any irreflexive, transitive relation over W. Then we say that

W is *well-founded by* $<$ iff every non-empty subset $A \subseteq W$ has at least one minimal element. This definition is rather compact, and needs to be unfolded carefully.

- A *minimal* element of a set $A \subseteq W$ (under $<$) is any $a \in A$ such that there is no $b \in A$ with $b < a$.

- The definition requires that *every* non-empty subset $A \subseteq W$ has a minimal element. The word 'every' is vital. It is not enough to find *some* subset A of W with a minimal element, nor to show that W itself has a minimal element. Thus the requirement is very demanding.

The definition of well-founding can be put in another, equivalent way. Let W be any set, and $<$ any irreflexive, transitive relation over W. Then W is well-founded by $<$ iff there is no *infinite descending chain* $\ldots a_3 < a_2 < a_1$ of elements of W.

To help appreciate the concept, we give some examples of well-founded sets.

- Clearly, the set $\mathbf{N}$ of all natural numbers is well-founded under the usual ordering, since every non-empty set of natural numbers has a minimal element (in fact a unique least element). This contrasts with the set $\mathbf{Z}$ of all integers, which is not well-founded under the customary ordering since it has a subset (e.g. the negative integers, or indeed the whole of $\mathbf{Z}$) that does not have a minimal element.

- A well-founded set can be 'longer' than $\mathbf{N}$. For example, if we take the natural numbers in their standard order, followed by the negative integers in the *converse* of their usual order, giving us the set $\{0, 1, 2, \ldots; -1, -2, -3, \ldots\}$, then this is well-founded in the order of listing. Clearly this can be repeated as many times as we like.

EXERCISE 4.7.1

Give an example of the same phenomenon but using the set $\mathbf{N}$ itself, suitably reordered.

- A well-founded set can also be 'wider' than $\mathbf{N}$. It need not be a *linear* (alias total) order: we may have elements a, b with neither $a = b$ nor $a < b$ nor $b < a$. For example, the collection of all *finite* subsets of an arbitrarily chosen set is well-founded under the relation of inclusion, even though when the chosen set has more than one element the collection will have two distinct subsets (e.g. two singletons) neither of which is properly included in the other.

EXERCISE 4.7.2

(a) Explain why, as claimed, the collection of all finite subsets of **N** is well-founded.

(b) Show that the collection of all subsets (finite or infinite) of **N** is not well-founded. *Hint*: Use the definition in terms of infinite descending chains.

• In the special case that a set is well-founded by a total relation, i.e. for all a,b in the set, either $a = b$ or $a < b$ or $b < a$, we say that it is *well-ordered*. Thus, a set W is well-ordered by a relation $<$ iff $<$ is a linear order of W satisfying the condition that every non-empty subset of W has a minimal element.

EXERCISE 4.7.3

(a) Show that if W is well-ordered then every non-empty subset $A \subseteq W$ has a *unique* minimal element, which is moreover *the least* element of the subset, in the sense that it is less than every other element of the subset.

(b) Show how, from any well-ordered set we may form one that is not well-ordered, but is still well-founded, by adding a single element.

4.7.2 The Principle of Proof by Well-Founded Induction

Roughly speaking, a well-founded relation provides a ladder up which we can climb in a set. This intuition is expressed rigorously by the *principle of induction over well-founded sets*, which may be formulated as follows. Let W be any well-founded set, and consider any property. To show that every element of W has the property, it suffices to show:

> *Induction step*: the property holds of an arbitrary element $a \in W$ whenever it holds of all $b \in W$ with $b < a$.

Note that the principle has no basis. In this it is like cumulative induction on the positive integers. Indeed, it may be seen as a direct abstraction of cumulative induction, from the specific order that we are familiar with there to any well-founded relation whatsoever. Of course, we could write in the basis, which would be: Every minimal element of W itself has the property in question. But it would be redundant, and in this context is customarily omitted.

EXERCISE 4.7.4 (WITH SOLUTION)

(a) Formulate the principle in contrapositive form.

(b) Formulate the principle as a statement about subsets of W rather than about properties of elements of A.

(c) Formulate the principle in contrapositive form as a statement about subsets of W.

Solution:

(a) Let W be any well-founded set, and consider any property. If the property does not hold of all elements of W, then there is an $a \in W$ that lacks the property although every $b \in W$ with $b < a$ has the property.

(b) Let W be any well-founded set, and let $P \subseteq W$. If $a \in P$ whenever $b \in P$ for every $b \in W$ with $b < a$, then $P = W$.

(c) Let W be any well-founded set, and let $P \subseteq A$. If $P \neq W$ then there is an $a \in W$ with $a \notin P$ although $b \in P$ for every $b \in W$ with $b < a$.

The proof of the principle of induction over well-founded sets is remarkably brief, considering its power. We use proof by contradiction. Let W be any well-founded set, and consider any property. Suppose that (1) the property holds of an arbitrary element $a \in W$ whenever it holds of all $b \in W$ with $b < a$, but (2) it does not hold of all elements of W. We get a contradiction.

Let A be the set consisting of those elements of W that do *not* have the property in question. By the second supposition, A is not empty so, since W is well-founded, A has a minimal element a. Thus on the one hand, a does not have the property. But on the other hand, since a is a minimal element of A, every $b \in W$ with $b < a$ is outside A, and so *does* have the property in question, contradicting supposition (1), and we are done.

Alice Box: proving the principle of induction over well-founded sets

Alice: That's certainly brief for such a general principle – although I would not say it is easy. But there is something about it that bothers me. We showed that the principle holds for any set that is well-founded under a relation $<$.

Hatter: Yes, indeed.

(*Continued*)

Alice Box: (Continued)

Alice: In the definition of a well-founded set, we required that the relation $<$ be irreflexive and transitive, as well as satisfying the 'minimal element' condition that every non-empty subset has at least one minimal element. I see how we used the minimal element condition in the proof, but I don't see that we used irreflexity or transitivity. Does this mean that we can generalize the principle by dropping those two requirements?

Hatter: Very perceptive! The short answer is 'yes'. A longer answer adds that the further generality is partly apparent, because the minimal element condition itself implies irreflexivity.

Alice: And transitivity?

Hatter: No, it does not imply transitivity – although it does imply acyclicity, defined in the exercises at the end of Chapter 2. So you can indeed get a bit more generality into the principle by dropping the transitivity assumption. But the present formulation covers most applications.

EXERCISE 4.7.5

> Verify the claims made by the Hatter: that the minimal element condition implies both irreflexivity and acyclicity.

4.7.3 Definition of a Function by Well-Founded Recursion on its Domain

Induction for well-founded sets is a principle of proof. Is there a corresponding principle for definition, i.e. guaranteeing the existence of functions that are defined by recursion on a well-founded domain? The answer is positive. However stating the principle in its full generality in a fully formal manner is quite subtle, and we will not attempt it here. We will give a rather informal formulation that covers most of the cases that a computer scientist is likely to need.

Principle of recursive definition on well-founded sets: Let W be any set that is well-founded under a relation $<$. Then we may safely define a function $f\colon W \to X$ by giving a rule that specifies, for every $a \in W$, the value of $f(a)$ in terms of the values of $f(b)$ for a collection of $b < a$, using any other functions and sets already well defined. 'Safely' here means that *there exists a unique function satisfying the definition*.

For confident and adventurous readers who have managed to follow so far, we illustrate the principle by using it to show that the Ackermann function is well-defined. Those not so confident may skip it, passing directly to the next section. We recall the recursive definition of the Ackerman function.

$$A(0, n) = n + 1$$
$$A(m, 0) = A(m - 1, 1) \text{ for } m > 0$$
$$A(m, n) = A(m - 1, A(m, n - 1)) \text{ for } m, n > 0$$

This function has two arguments, both from $\mathbf{N}$, so we turn it into a one-argument function on $\mathbf{N^2} = \mathbf{N} \times \mathbf{N}$ by reading the round brackets in the definition as indicating ordered pairs (strictly speaking, we should then add extra round brackets for redundant punctuation, but we keep it reader-friendly). Given the familiar relation $<$ over $\mathbf{N}$, which we know to be well-founded, indeed to be a well ordering of $\mathbf{N}$, we define the *lexicographic order* over $\mathbf{N^2}$ by the rule: $(m,n) < (m',n')$ iff either $m < m'$, or $m = m'$ and $n < n'$.

This is a very useful way of ordering the Cartesian product of any two well-founded sets, and deserves to be remembered in its own right. The reason for the name 'lexicographic' will be clear if you think of the case where instead of $\mathbf{N}$, we are considering its initial segment $A = \{n : 0 \leq n \leq 25\}$ and identify these with the letters of the English alphabet. The lexicographic order of A^2 then corresponds to its dictionary order.

We now check that the lexicographic relation $<$ is irreflexive, transitive, and that it well-founds $\mathbf{N^2}$ (in fact, it is also linear and so well-ordered). Having done this, we are ready to apply the principle of recursive definition on well-founded sets. All we have to do is show that the two recursion clauses of the candidate definition specify, for every $(m,n) \in \mathbf{N^2}$, the value of $A(m,n)$ in terms of the values of $A(p,q)$ for a collection of pairs $(p,q) < (m,n)$, using any other functions and sets already known to exist.

This amounts to showing, for the first clause of the recursion step, that $(m{-}1,1) < (m,0)$ and, for the second clause of the recursion step, that $(m{-}1, A(m, n{-}1)) < (m,n)$, for all $m, n > 0$. But both of these are immediate from the definition of the lexicographic order $<$. We have $(m{-}1,1) < (m,0)$ because $m{-}1 < m$, regardless of the fact that $1 > 0$. Likewise, we have $(m{-}1, A(m, n{-}1)) < (m,n)$ no matter how large $A(m, n{-}1)$ is, again simply because $m{-}1 < m$.

EXERCISE 4.7.6

Check the claim made, that the lexicographic order of $\mathbf{N^2}$ is irreflexive, transitive, and that it is well-founds $\mathbf{N^2}$, indeed, is also linear and so well-ordered.

4.8 Recursive Programs

What is the link between all this and *recursive programs* in computer science? Does the mathematical work on recursive definitions and inductive proofs do anything to help us understand how programs work and what they can do?

The link lies in the fact that a program can be seen as a finite battery of instructions to perform computations for a potentially infinite range of inputs. Simplifying a great deal, and restricting ourselves to the principle case of *deterministic* programs (those in which the instructions suffice to fully determine each step in terms of its predecessors) the program may be understood as a recipe for defining a function that takes an input a to a finite or infinite succession a_0, a_1,... of computed items. If the sequence is finite, the last item may be regarded as the output.

Equivalently, we are looking at a recipe for defining a two-place function taking each pair (a,n), where a is a possible input and n is a natural number, to an item a_n. The recipe is recursive in the sense that the value of a_n may depend upon the values of any of the earlier a_m, in whatever way that the program specifies, so long as each step to a_n from the earlier steps may actually be performed by the computer that is executing the program. Typically, the 'while' and 'until' clauses setting up loops in the program, correspond to recursion clauses in the definition of the associated function.

The question thus arises: When does the proposed recipe really give us a program, i.e. when does it really define a function? If we are writing the program in a well-constructed programming language, then the tight constraints that have been imposed on the grammar of the language may suffice to guarantee that it is well-defined. But if we are working in a lax or rather informal language, the question of whether we have specified a unique function needs to be analysed using the concept of recursive definition.

When the recipe does give us a unique function, other questions remain. Is the program guaranteed to terminate, i.e. is the sequence of steps a_n finite? If it does terminate, does the output give us what we would like it to give? Answers to these questions require *proof*, and the proof will always involve inductive arguments of one kind or other, perhaps of many kinds. In some simple cases it can be sufficient to induce on the number of loops in the program. For more complex programs, the inductive arguments can be less straightforward. In general, one needs to devise a suitable relation over a suitable set of items, verify that it is well founded, and then use well-founded induction over the relation.

FURTHER EXERCISES

4.1. *Proof by simple induction*

(a) Use simple induction to show that for every positive integer n, 5^n-1 is divisible by 4.

(b) Use simple induction to show that for every positive integer n, n^3-n is divisible by 3. *Hint*: In the induction step, you will need to make use of the arithmetic fact that $(k+1)^3 = k^3 + 3k^2 + 3k + 1$.

4.2. *Definition by simple recursion*

(a) Let $f: \mathbf{N} \to \mathbf{N}$ be the function defined by putting $f(0) = 0$ and $f(n+1) = n$ for all $n \in N$. (i) Evaluate this function bottom-up for all arguments 0–5. (ii) Explain what f does by expressing it in explicit terms (i.e. without a recursion). *Hint*: Your explanation will make use of subtraction.

(b) Let $g: \mathbf{N} \times \mathbf{N} \to \mathbf{N}$ be defined by putting $g(m,0) = m$ for all $m \in N$ and $g(m,n+1) = f(g(m,n))$ where f is the function defined in the first part of this exercise. (i) Evaluate $f(3,4)$ top-down. (ii) Explain what f does by expressing it in explicit terms (i.e. without a recursion). *Hint*: Your explanation will again make use of subtraction.

(c) Let $f: \mathbf{N}^+ \to \mathbf{N}$ be the function that takes each positive integer n to the greatest natural number p with $2^p \leq n$. Define this function by a simple recursion. *Hint*: You will need to divide the induction step into two cases.

4.3. *Proof by cumulative induction*

(a) Use cumulative induction to show that any postage cost of four or more pence can be covered by two-pence and five-pence stamps.

(b) Use cumulative induction to show that for every natural number n, $F(n) \leq 2^n-1$, where F is the Fibonacci function.

(c) Calculate $F(5)$ top-down, and then again bottom up, where F is the Fibonacci function.

(d) Express each of the numbers 14, 15, and 16 as a sum of 3s and/or 8s. Using this fact in your basis, show by cumulative induction that every positive integer $n \geq 14$ may be expressed as a sum of 3s and/or 8s.

(e) Show by induction that for every natural number n, $A(1,n) = n + 2$, where A is the Ackermann function.

4.4. Definition of sets by structural recursion

(a) Define by structural recursion the set of all odd palindromes over a given alphabet.

(b) Define by structural recursion the set of all palindromes (over a given alphabet) that contain no two contiguous occurrences of the same letter. *Hint*: Make use of the fact that you will verify in the first part of the next exercise.

4.5. Proof by structural induction

(a) Show by structural induction that every even palindrome is either empty or contains two contiguous occurrences of the same letter.

(b) Show by structural induction that in classical propositional logic, no formula built using propositional letters and connectives from among $\vee$, $\wedge$, $\rightarrow$, $\leftrightarrow$ is a contradiction. *Hint*: Think of a suitable assignment of truth-values, and show by structural induction that it makes every such formula true.

(c) Justify the principle of structural induction by using cumulative induction over the natural numbers. *Hint*: To keep the exposition simple, consider the closure A^+ of A under a single relation R, and use the alternative definition of A^+ as the union of sets A_0, $A_1, \ldots$

4.6. Definition of functions by structural recursion on their domains

(a) Consider any alphabet A and the set A^* of all strings made up from A. Intuitively, the reversal of a string s is the string obtained by reading all its letters from right to left instead of left to right. Provide a structural recursive definition for this operation on strings. *Hint*: The base should concern the empty string, called λ, and the induction step should use the operation *cons*.

(b) In presentations of propositional logic, it is customary to define an *assignment* of truth values to be any function $v: A \rightarrow \{1,0\}$, where A is some stock of elementary letters p, q, $r, \ldots$ assumed to be given. Any such assignment is then extended to a valuation $v^+: A^+ \rightarrow \{1,0\}$ where A^+ is the set of all formulae, i.e. the closure of A under the operations of forming, say, negations, conjunctions and disjunctions, in a way that respects the usual truth-tables. It is tacitly assumed that this extension is unique. Analyse what is being done in terms of structural recursion.

4.7. *Recursion and induction on well-founded sets*

(a) Prove the equivalence of the two definitions of a well-founded set.

(b) Show that every subset of a well-founded set is well-founded under the same relation.

(c) Is the converse of a well-founding relation always a well-founding relation? Give proof or counter-example.

Selected Reading

Induction and recursion on the positive integers. There are plenty of texts, though most tend to neglect the recursion side in favour of induction. Here are two among others:

> Carol Schumacher *Chapter Zero*: *Fundamental Notions of Abstract Mathematics*. Pearson, 2001 (second edition), Chapter 3.

> James L. Hein *Discrete Structures, Logic and Computability*. Jones and Bartlett, 2005 (second edition), Chapter 4.4.

Well-founded induction and recursion. Introductory accounts tend to be written for students of mathematics rather than computer science, and again tend to neglect recursive definition. Two texts accessible to computer science students:

> Seymour Lipschutz *Set Theory and Related Topics*. McGraw Hill Schaum's Outline Series, 1998, Chapters 8–9.

> Paul R. Halmos *Naive Set Theory*. Springer, 2001 (new edition), Chapters 17–19.

Structural induction and recursion. It is rather difficult to find introductory presentations. One of the purposes of the present chapter has been to fill the gap.

5

Counting Things: Combinatorics

Chapter Outline

Up to now, our work has been almost entirely qualitative. The concepts of a set, relation, function, recursion and induction are non-numerical, although they have important numerical manifestations as, say, sets of integers or recursive definitions on the natural numbers. In this chapter we turn to quantitative matters, and specifically to problems of counting. We will tackle two kinds of question.

First: Are there rules for determining the number of elements of a large set from the number of elements of smaller ones? Here we will learn how to use two very simple rules: *addition* (for unions of disjoint sets) and *multiplication* (for Cartesian products of arbitrary sets).

Second: Are there rules for calculating the number of possible selections of k items out of a set with n elements? Here we will see that the question is less straightforward, as there are several different kinds of selection, giving us very different outcomes. We will untangle *four basic modes of selection*, give arithmetical formulae for them, and practice their application. In the final section we will turn to the problems of counting *rearrangements* and *partitions* of a set.

D. Makinson, *Sets, Logic and Maths for Computing*,
DOI: 10.1007/978-1-84628-845-6_5, © Springer-Verlag London Limited 2008

5.1 Two Basic Principles: Addition and Multiplication

In earlier chapters, we already saw some principles for calculating the number of elements of one set, given the number in another. In particular, in Chapter 3 we noted the *equinumerosity principle*: two finite sets have the same number of elements iff there is a bijection from one to the other. In the exercises at the end of Chapter 1, we noted an important equality for difference, and another one for disjoint union. They provide our starting point in this chapter, and we begin by recalling them. For difference we observed:

> *Subtraction principle for difference*: Let A, B be finite sets.
> Then $\#(A \backslash B) = \#(A) - \#(A \cap B)$.

For union we saw:

> *Addition principle for two disjoint sets*: Let A, B be disjoint finite sets.
> Then $\#(A \cup B) = \#(A) + \#(B)$.

The condition of disjointedness is essential here. For example, if $A = \{1,2\}$ and $B = \{2,3\}$ then $A \cup B = \{1,2,3\}$ with only three elements, not four.

Clearly the addition principle can be generalized to n sets, provided they are *pairwise disjoint* in the sense of Chapter 1, i.e. for any $i, j \leq n$, if $i \neq j$ then $A_i \cap A_j = \varnothing$.

> *Addition principle for many pairwise disjoint sets*: Let $A_1, \ldots, A_n$ be
> pairwise disjoint finite sets Then $\#(\cup \{A_i\}_{i \leq n}) = \#(A_1) + \ldots + \#(A_n)$.

EXERCISE 5.1.1 (WITH SOLUTION)

> Let A, B be finite sets. Formulate and verify a necessary and sufficient condition for $\#(A \cup B) = \#(A)$.
>
> *Solution:* $\#(A \cup B) = \#(A)$ iff $B \subseteq A$. Verification: If $B \subseteq A$ then $A \cup B = A$ so $\#(A \cup B) = \#(A)$. Conversely, if the inclusion does not hold then there is a $b \in B \backslash A$ so $\#(A) < \#(A) + 1 = \#(A \cup \{b\}) \leq \#(A \cup B)$.

Can we say anything for union when the sets A, B are not disjoint? Yes, by breaking them down into disjoint components. For example, we know that

$A \cup B = (A \cap B) \cup (A \backslash B) \cup (B \backslash A)$ and these are disjoint, so $\#(A \cup B) = \#(A \cap B) + \#(A \backslash B) + \#(B \backslash A)$. Hence by the subtraction principle for difference;

$$\#(A \cup B) = \#(A \cap B) + \#(A) - \#(A \cap B) + \#(B) - \#(A \cap B)$$
$$= \#(A \cap B) - \#(A \cap B) - \#(A \cap B) + \#(A) + \#(B)$$
$$= \#(A) + \#(B) - \#(A \cap B).$$

We have thus shown the:

Addition Principle for Two Arbitrary Sets : Let A, B be any finite sets. Then $\#(A \cup B) = \#(A) + \#(B) - \#(A \cap B)$.

EXERCISE 5.1.2

(a) Illustrate the addition principle for two disjoint sets using a Venn diagram.

(b) Do the same for the non-disjoint version.

(c) Check the claim that $A \cup B = (A \cap B) \cup (A \backslash B) \cup (B \backslash A)$.

(d) Show that $A \backslash B = A \backslash A \cap B$.

Alice Box: Addition principle for n arbitrary sets

Alice: We generalized the principle for two disjoint sets to n disjoint sets. Can we make a similar generalization from two arbitrary sets to n arbitrary sets?

Hatter: Indeed we can, but its formulation is rather more complex. To state it properly we need the notion of an arbitrary combination of n things k at a time, which we will get to later in the chapter. So let's take a rain-check on that one.

EXERCISE 5.1.3 (WITH SOLUTION)

(a) The classroom contains seventeen male students, eighteen female students, and the professor. How many altogether in the classroom?

(b) The logic class has twenty students who also take calculus, nine who also take a philosophy unit, eleven who take neither, and two who take both calculus and a philosophy unit. How many students in the class?

Solution:

(a) 17+18+1 = 36, using the addition principle for $n = 3$ disjoint sets.

(b) ((20+9)–2)+11 = 38, using the addition principle for two arbitrary sets to calculate the number taking either calculus or philosophy (in the outer parentheses), and then applying the addition principle for two disjoint sets to cover those who take neither.

Applications of the addition principles taken alone are evidently quite trivial. Their power comes from their joint use with other rules, notably the multiplication principle, which we now explain:

Multiplication rule for two sets : Let A, B be finite sets. Then

$$\#(A \times B) = \#(A) \cdot \#(B).$$

In words: the number of elements in the Cartesian product of two sets is equal to the product of the numbers in each. Reason: If A has m elements and B has n elements, then for each of the m elements of the set A serving as a first term in an ordered pair $(a,b) \in A \times B$, there are n elements of B that can serve as second term b: the selection we make from A does not limit or influence the range of subsequent selection in B. The same reasoning gives us more generally:

Multiplication rule for many sets. Let $A_1, \ldots, A_n$ be finite sets. Then

$$\#(A_1 \times \ldots \times A_n) = \#(A) \cdot \ldots \cdot \#(B).$$

EXERCISE 5.1.4 (WITH SOLUTION)

The menu at our restaurant allows choice: for first course one can choose either soup or a salad, the second course can be either beef, duck or vegetarian, followed by a choice of fruit or cheese, ending with black tea, green tea or coffee. How many selections are possible?

Solution: $2 \cdot 3 \cdot 2 \cdot 3 = 36$. The selections are independent of each other, and so are in one-one correspondence with the Cartesian product of four sets with 2,3,2,3 elements respectively.

Alice Box: The maverick diner

Alice: Not if I am there!

Hatter: What do you mean?

Alice: Well, I never drink anything with caffeine in it, so I would skip the last choice. And following an old European custom, I prefer to take my fruit before anything else, so I would not follow the standard order. So *my* selection would not be any of your 36. It would be, say, (fruit, salad, duck, nothing).

Hatter: You have put your finger on an important issue. When a 'real-life' counting problem is given, its presentation is frequently underdetermined. In other words, certain aspects are left implicit, and they can often be filled in different ways. For example, in the restaurant problem it was tacitly assumed that everyone chooses exactly one dish from each category, and that the order of the categories is fixed: any other reordering is either disallowed or regarded as equivalent to the listed order. If we do not make these assumptions, we can get quite different numbers of selections.

Alice: That seems to make the mathematics difficult.

Hatter: Not so much the mathematics, but its application. In general, the trickiest part of a real-life counting problem is not to be found in the calculations to be made when applying a standard mathematical formula. It lies in working out *which* (if any) of the mathematical formulae in your tool-kit is the appropriate one. And that depends on understanding what the problem is about and locating its possible nuances. We need to know what items are being selected, from what categories, in what manner. Especially, we need to know when two items, or two categories, or two selections are to be regarded as identical – to be counted as one, not as two. Only then can we safely give the problem an abstract representation that justifies the application of one of the formulae in the tool kit. We will see more examples of this as we go on.

EXERCISE 5.1.5

(a) Telephone numbers in the London area begin with 020 and continue with eight more digits. How many are there?

(b) How many distinct licence plates are there consisting of two letters other than *O*, *I* and *S* (to prevent possible visual confusion with similar-looking digits), followed by four digits?

5.2 Using the Two Basic Principles Together

Often the solution of a problem requires the use of both the addition and multi-plication principles. Here is a simple example. How many four-digit numbers begin with 5 or with 73?

We begin by breaking our set (of all four-digit numbers beginning with 5 or with 73) into two disjoint subsets – those beginning with 5, and those beginning with 73. We determine their numbers separately, and then by the addition prin-ciple for disjoint sets we add them. In the first case, with the digit 5, there are three digits still to be filled in, with 10 choices for each, so we get $10^3 = 1000$ possibilities. In the second case, beginning with 73, we have two digits to fill in, hence $10^2 = 100$ possibilities. So the total is 1100 four-digit numbers beginning with 5 or with 73.

This kind of procedure is quite common, and we look at its general form. We are required to determine $\#(A)$ – the number of elements of a set A. We observe that $A = A_1 \cup ... \cup A_n$ where the sets A_i are pairwise disjoint. We then note that each of the sets A_i is (or may be put in one-one correspondence with) a Cartesian product $A_{i1} \times ... \times A_{ik}$ (where k may depend on i). So we apply n times the multi-plication rule for many sets to get the value of each $\#(A_i)$, and then apply once the addition rule for many disjoint sets to get the value of $\#(A)$ itself.

EXERCISE 5.2.1 (WITH PARTIAL SOLUTION)

(a) You have six shirts, five ties, and four pairs of jeans. You must wear a shirt and a pair of trousers, but maybe not a tie. How many outfits are possible?

(b) A tag consists of a sequence of four alphanumeric signs (letters or digits). How many tags with alternating letters and digits begin with either the digit 9 or the letter m? *Warning*: Pay attention to the requirement of alternation.

Solution to (a): The set of possible outfits is the union of two sets: $S \times J$ and $S \times J \times T$. Notice that these two sets are disjoint (why?). There are $6 \cdot 4 = 24$ elements of $S \times J$, and $6 \cdot 4 \cdot 5 = 120$ elements of $S \times J \times T$. So there are 144 attires in which to sally forth.

5.3 Four Ways of Selecting k Items out of n

A club has 20 members, and volunteers are needed to set up a weekly roster of members to clean up the coffee room, one for each of the six days of the week that the club is open. How many possible ways of doing this?

This question has no answer! Or rather, it has several different answers, according to how we interpret it. The general form is: how many ways of selecting k items (here 6 volunteers) out of a set with n elements (here, the pool of 20 members). Without looking for more far-fetched readings, there are two basic dimensions on which more precision is needed.

- Does the *order* in which the selection is made matter for counting purposes? For example, is the roster with me serving on Monday and you on Tuesday regarded as different from the one that is the same for all other days but with our time-slots swapped? Do we count them as two rosters or as one?

- Is *repetition* allowed? For example, are you allowed to volunteer for more than one day, say both Monday and Friday?

These two options evidently give rise to four cases, which we will need to distinguish carefully in our analysis.

Order matters, repetition allowed	O+R+
Order matters, repetition not allowed	O+R−
Order doesn't matter, repetition allowed	O−R+
Order doesn't matter, repetition not allowed	O−R−

To illustrate the difference graphically in the two dimensions that a page has to offer, suppose that we need only 2 volunteers out of, say, 5 members.

Table 5.1 Modes of selecting 2 items out of 5.

	a	b	c	d	e
a					
b					
c					
d					
e					

The members are called a,b,c,d,e. If we select volunteers (b,d) say, then this is represented by the cell for row b and column d. The table contains 25 cells altogether.

- **O+R+**: If order matters and repetition is allowed, then each cell represents a possible selection of two volunteers out of five. There are thus 25 possible selections.

- **O+R−**: If order matters but repetition is not allowed, then the diagonal cells do not represent allowed selections. There are thus $25-5 = 20$ possible selections.

- **O−R+**: If order does not matter but repetition is allowed, then the cells to the upper right of the diagonal give the same outcome as their images in the bottom left. For example, (b,d) represents the same selection as (d,b). In this case we have only $25-10 = 15$ possible selections, subtracting the ones to the bottom left of the diagonal to avoid counting them twice.

- **O–R–**: If order does not matter and repetition is not allowed, we have even less. We must subtract the cells of the diagonal from the previous count, leaving only those to the upper right. Thus we get only $15-5 = 10$ selections.

Thus, according to the way in which we understand the selection we get 25, 20, 15, or 10 possible selections! We must consider the four modes of selection one by one.

EXERCISE 5.3.1

(a) Identify the cells corresponding to O+R–, O–R+, O–R– by suitable hatching (vertical, horizontal, diagonal respectively).

(b) Construct a similar table for the four different modes of choosing 2 elements out of 6, hatching the respective areas in the same way, and giving the number of selections under each mode.

It will be helpful to note the following connections with what we have learned about functions in Chapter 3.

O+R+: When order matters and repetition is allowed, then each selection of k items from an n-element set may be represented by an ordered k-tuple $(a_1,\ldots,a_k)$ of elements of the set. As we already know, such an ordered k-tuple may be understood as a function on the set $\{1,\ldots,k\}$ into the set $\{1,\ldots,n\}$. Thus the number of selections of k items from an n-element set, understood in this mode, is the same as the number of *functions* from a k-element set e.g. $\{1,\ldots,k\}$ into an n-element set e.g. $\{1,\ldots,n\}$.

O+R–: When order matters but repetition is not allowed, then the number of selections of k items from an n-element set is the same as the number of *injective functions* from a k-element set e.g. $\{1,\ldots,k\}$ into an n-element set e.g. $\{1,\ldots,n\}$. Injective, because when $i \neq j$, $f(i)$ is not allowed to be the same as $f(j)$.

O–R–: When order does not matter and repetition is not allowed, then the number of selections of k items from an n element set is the same as the number of *ranges of injective functions* from a k-element set into an n-element set. If you think about it, this is the number of k-element *subsets* of the n-element set. Reason: every such range is a k-element subset of the target set, and every k-element subset of the target set is the range of some such function; the two collections are thus identical and so have the same number of elements.

Alice Box: The mode O–R+, sets and multisets

Alice: You forgot the other mode O–R+ of selection – where order doesn't matter but repetition is allowed. What does it amount to in terms of functions?l

Hatter: Actually, I did not forget. I deliberately omitted it.

Alice: Why?

Hatter: It is not so simple, and may confuse you.

Alice: Tell me anyway.

Hatter: One way of describing it is as the number of functions f from a set $\{a_1,\ldots,a_n\}$ with n elements into $\{0,\ldots,k\}$ such that $f(a_1)+\ldots+f(a_n) = k$.

Alice: I don't get it.

Hatter: The function tells us how many times each element a_i appears in the selection (zero or more), and the sum of these appearances must come to k.

Alice: So we are counting functions again?

Hatter: Yes, but be careful with the domain of these functions, which this time has n elements, and the target set, which has $k+1$ elements. There is also another way of understanding such selections, in a language we have not seen before. We can see them as the k-element *multisets* that can be formed from a *set* of n elements.

Alice: What is a multiset?

Hatter: Roughly speaking, it is like a set, but allows repetitions. Thus when a and b are distinct items, the *set* $\{a,a,b,b\}$ has only *two* elements and is identical to the set $\{a,b\}$, whereas the multiset consisting of a, a, b, b has *four* members, and is distinct from the multiset consisting of just a, b. Sometimes also called bags, multisets have recently gained some popularity in certain areas of logic and computer science. But in my view, in an introductory account like this it is better to stay with the language of sets.

Alice: OK, let's forget multisets, and think in terms of sets and functions. How do you *write* the selections made under the mode O–R+? I mean, under the two modes O+R± where order is significant we get ordered k-tuples and we write them as $(a_1,\ldots,a_k)$; under the mode O–R– we get subsets, so we write them as $\{a_1,\ldots,a_k\}$; but what about selections made under the mode O–R+?

(Continued)

Alice Box: (Continued)

Hatter: Good question! Of course, if you are thinking of them as multisets, it is natural to write them with some other kind of bracket, say square ones. But from our perspective, thinking of them as functions, it is better to write them as sets of ordered pairs. So if we make k-selections from a set $\{a_1,\ldots,a_n\}$ in this mode, each such selection will be written as an n-element set $\{(a_1,i_1),\ldots,(a_n,i_n)\}$ of ordered pairs, where each $0 \leq i_j \leq k$ and $i_1+\ldots+i_n = k$.

EXERCISE 5.3.2 (WITH SOLUTION)

When $k > n$, which of the four modes of selection become impossible?

Solution: The two modes that disallow repetition (O+R– and O–R–) prevent any possible selection when $k > n$. For they require that there be at least k *distinct* items selected, which is not possible when there are less than k waiting to be chosen.

On the other hand, the two modes permitting repetition (O+R+ and O–R+) still allow selection when $k > n$. For example the ordered triple (a,b,a) represents a selection of three items under mode O+R+, out of a set $\{a,b\}$ of two items. Likewise, using the notation introduced in the last Alice box, the (unordered) set $\{(a,2), (b,1)\}$ represents a selection of three items out of the same set $\{a,b\}$ under mode O–R+.

So far, we have been calling the four modes by their codes; it is time to give them names. Two of the four modes – those where repetition is *not* allowed – have long-standing and universally accepted names.

- When order matters (and repetition is not allowed), i.e. O+R–, the mode of selection is called *permutation*. The function counting the number of permutations of n elements k at a time is written $P(n,k)$, or in some texts as nPk or ^nP_k.

- When order does not matter (and repetition is not allowed), i.e. O–R–, the mode of selection is called a *combination*. The function counting the number of combinations of n elements k at a time is written $C(n,k)$, or for some authors, nCk or ^nC_k. A common two-dimensional notation, very convenient in complex calculations, puts the n above the k within large round brackets.

These two modes are the ones that tend to crop up most frequently in traditional mathematical practice, e.g. in the formulation of the binomial theorem, which goes back to the sixteenth century. They are also the most common in computer science practice. On the other hand, for the modes allowing repetition, terminology is distressingly variable from author to author. The simplest names for these two modes piggy-back on the standard ones for the modes without repetition, as follows.

- When order matters (but now repetition is allowed), i.e. when we are in the mode O+R+, we will speak of *permutations with repetition allowed.*

- When order does not matter (but now repetition is allowed), i.e. when we are in the mode O–R+, we will speak of *combinations with repetition allowed.*

In the following table we summarize the nomenclature. Note that it lists the four modes in a sequence different from that which was convenient for a conceptual explanation. The order in Table 5.2 is more convenient for numerical analysis; it corresponds to the order in which we will shortly establish their respective counting formulae. It also corresponds, very roughly, to their relative importance in applications.

We use the term *selection* to cover, quite generally, all four modes. The subject of counting selections is often nicknamed the theory of *perms and coms.*

Table 5.2 Four modes of selecting k items from an n-element set

Generic Term	Mode of Selection	Particular Modes	Notation
	O+R–	Permutations	$P(n,k)$
	O–R–	Combinations	$C(n,k)$
Selection	O+R+	Permutations with repetition allowed	
	O–R+	Combinations with repetition allowed	

5.4 Counting Formulae: Permutations and Combinations

So far we have been doing essentially conceptual work, sorting out different modes of selection. We now present counting formulae for the four modes of selection that were distinguished. We begin with the two in which repetition is not allowed: permutations and combinations.

Table 5.3 Formulae for k-permutations and k-combinations (without repetition) from an n-element set.

Mode of Selection	Notation	Standard Name	Formula	Proviso
O+R–	$P(n,k)$	Permutations	$n! \,/\, (n{-}k)!$	$k \leq n$
O–R–	$C(n,k)$	Combinations	$n! \,/\, k!(n{-}k)!$	$k \leq n$

At this stage, it is tempting to plunge into a pool of examples to practice applying the counting formulae. Indeed, this is necessary if one is to master the material. But by itself it is not enough. We also need to see *why* each of the formulae does its job; that also helps understand *when* it, rather than its neighbour, should be used in a given problem.

EXERCISE 5.4.1

(a) Check that the figures obtained in Section 5.3 for choosing 2 out of 5 agree with these formulae.

(b) Calculate the figures for choosing 4 out of 9.

5.4.1 The Formula for Permutations (O+R−)

For intuition, we begin with an example. How many six-digit numerals are there in which no digit appears more than once? The formula in our table says that it is 10! / (10−6)! = 10! / 4! = $10 \cdot 9 \cdot 8 \cdot 7 \cdot 6 \cdot 5$ = 151,200. How do we get this?

We use the multiplication principle. There are n ways of choosing the first item. As repeated choice of the same element is *not* allowed, we thus have only n−1 ways of choosing the second, then only n−2 ways of choosing the third, and so on. If we do this k times, the number of possible selections is thus: $n \cdot (n{-}1) \cdot (n{-}2) \cdot \ldots \cdot (n{-}(k{-}1))$. Multiplying this by $(n{-}k)!$ / $(n{-}k)!$ gives us $n!$ / $(n{-}k)!$ in agreement with the counting formula.

Alice Box: The proviso $k \leq n$

Alice: Why the condition that $k \leq n$?

Hatter: Well, conceptually, as we noticed in an exercise earlier in this chapter, it is impossible to select more than n elements from an n-element set if we do not allow repetition. Numerically, when $k > n$ then $n{-}k$ and so also $n!$ / $(n{-}k)!$ are negative, which doesn't make much sense in counting. So we leave the function $C(n,k)$ undefined when $k > n$ or, if you prefer, set it at 0.

Alice: OK. But what happens when $k = n$? It worries me, because in that case $n{-}k$ is 0, and in Chapter 4 the factorial function was defined only for *positive* integers.

(Continued)

> *Alice Box:* (Continued)
>
> *Hatter:* True. To cover that case, it is conventional to extend the definition of factorial by setting $0! = 1$. So when $k = n$ we have $P(n,k) = P(n,n) = n! / (n-n)! = n! / 0! = n!$
>
> *Alice:* So we can say that $P(n,n) = n!$
>
> *Hatter:* Yes, and that is an important case of the general principle. We will be using it very shortly.

EXERCISE 5.4.2 (WITH PARTIAL SOLUTION)

(a) Use the formula to calculate each of $P(6,0),\ldots P(6,6)$.

(b) You have 15 ties, and you want to wear a different one each day of the working week (five days). For how long can you go on without ever repeating a weekly sequence?

Solution to (b): Our base set is the set of ties, with 15 elements, and we are selecting 5 with order significant and repetition not allowed. We can thus apply the formula for permutations. There are thus $15! / (15-5)! = 15! / 10! = 15 \cdot 14 \cdot 13 \cdot 12 \cdot 11 = 360{,}360$ possible weekly selections. That means you can go on for nearly 7,000 years without repeating a weekly sequence. Moral: you have too many ties!

EXERCISE 5.4.3 (WITH SOLUTION)

(a) How would you interpret the meaning of $P(n,k)$ when $k = 0$, in words of ordinary English? Is the counting formula reasonable in this limiting case?

(b) Compare the values of $P(n,n)$ and $P(n,0)$. Any comments?

Solutions:

(a) $P(n,0)$ is the number of ways of selecting nothing from a set of n elements. The counting formula gives the value $P(n,0) = n! / n! = 1$. That is, one way of selecting nothing.

(b) $P(n,n) = n! / (n-n)! = n! / 0! = n!$ while as we have seen, $P(n,0) = 1$. Clearly $P(n,k)$ takes its largest possible value when $k = n$, and its smallest when $k = 0$.

5.4.2 The Formula for Combinations (O–R–)

For concreteness, we again begin with an example. How many ways are there of choosing a six-person subcommittee out of a full ten-person committee? It is understood here that the order of the members of the subcommittee is of no consequence, and that all six members of the subcommittee must be different people. That is, we are counting *combinations* (mode O–R–). The counting formula in the table gives us $C(10,6) = 10! / 6!4! = 10 \cdot 9 \cdot 8 \cdot 7 / 4 \cdot 3 \cdot 2 = 5 \cdot 3 \cdot 2 \cdot 7 = 210$ – a tiny fraction of the 151,200 for permutations (to be precise, one 6!th i.e. one 720th). Order matters!

To understand the reasoning behind this formula, notice that it says in effect that $C(n,k) = P(n,k) / k!$, so it suffices to prove that. Recall that $C(n,k)$ counts the number of k-element subsets of an n-element set. We already know that each such subset can be given $P(k,k) = k!$ orderings. Hence the total number $P(n,k)$ of *ordered* subsets is $C(n,k) \cdot k!$ so that $C(n,k) = P(n,k) / k! = n! / k!(n-k)!$.

EXERCISE 5.4.4 (WITH PARTIAL SOLUTION AND COMMENTS)

(a) Use the formula to calculate each of $C(6,0),\ldots,C(6,6)$.

(b) Draw a chart with the seven values of k ($0 \leq k \leq 6$) on the abscissa (horizontal axis) and the values of each of $P(6,k)$ and $C(6,k)$ on the ordinate (vertical axis).

(c) Your investment advisor has given you a list of eight stocks attractive for investment. You decide to invest in three of them. How many different selections are possible?

(d) Same scenario, except that you decide to invest $1,000 in one, double that in another, and double that again in a third. How many different selections are possible?

(e) Same scenario, except that your adviser also gives you a ranking of the stocks by risk. You decide to invest $1,000 in a risky one, double that in a less risky one, and double that again in an even less risky one. How many different selections are possible?

Solutions to (c) and (d) and comments on (e): These three questions illustrate how important it is to understand, before any calculation, what kind of selection we are supposed to be making.

(c) It is implicit in the formulation of this question that you wish to invest in three *different* stocks – no repetition. It is also implicit that the order of the selection is to be disregarded – we are interested only in the subset selected. So we are counting combinations (mode O–R–) and can apply the formula for $C(n,k)$, to get $C(8,3) = 8! \,/\, 3!5! = 8 \cdot 7 \cdot 6 \,/\, 3 \cdot 2 = 4 \cdot 7 \cdot 2 = 56$.

(d) In this question it is again assumed that there is no repetition, but the order of the selection is regarded as important – we are interested in which stock is bought in what volume. So we are back with permutations (mode O+R–), and should apply the counting formula for $P(n,k)$, to get $P(8,3) = 8! \,/\, 5! = 8 \cdot 7 \cdot 6 = 336$.

(e) *Comments*: This question is much trickier, and makes quite a few implicit assumptions. It assumes that the ranking of estimated risk does not put all stocks on the same level, indeed that there are at least three different levels. To simplify life, let's suppose that the risk ranking is linear – no two stocks on the same level. So we have 8 levels.

Now, the first choice must not be from the least risk level, nor from next risk level, but may be from any of the subsequent risk levels. The second choice depends to a certain extent on the first one: it must be on a lesser risk level, but still not the least risky one. And the third choice depends similarly on the second. We thus have quite a complex problem – more than a simple application of any one of the counting formulae in our table. A moral of this story is that real-life counting problems, apparently simple in their verbal presentation, can be quite nasty to solve.

EXERCISE 5.4.5 (WITH PARTIAL SOLUTION)

(a) How would you interpret the meaning of $C(n,k)$ when $k = 0$, in words of ordinary English?

(b) Why is $C(n,k) \le P(n,k)$ for all n,k with $1 \le k \le n$?

(c) State and prove a necessary and sufficient condition on n,k for $C(n,k) < P(n,k)$.

(d) Compare the values of $C(n,n)$, $C(n,1)$ and $C(n,0)$ with their counterparts $P(n,n)$, $P(n,1)$ and $P(n,0)$ and explain the similarities/differences in intuitive terms.

(e) Show from the counting formula that $C(n,k) = C(n,n{-}k)$.

(f) Suppose that n is even, i.e. $n = 2m$. Show that for $j < k \le m$ we have $C(n,j) < C(n,k)$.

(g) Same question but with n odd, i.e. $n = 2m{+}1$.

Solutions to (d), (e):

(d) $C(n,n) = n! \,/\, n!(n{-}n)! = n! \,/\, n!0! = 1$, while $C(n,1) = n! \,/\, 1!(n{-}1)! = n! \,/\, (n{-}1)! = n$, and $C(n,0) = n! \,/\, 0!(n{-}0)! = n! \,/\, 0!n! = 1$. Compare this with $P(n,n) = n!$, $P(n,1) = n$ and $P(n,0) = 1$ as already established.

Explanation: We have $C(n,0) = P(n,0)$ because there is just one way of selecting nothing, irrespective of order. Also, $C(n,1) = P(n,1)$ because when you are selecting just one thing, the question of order of selection does not arise. But $C(n,n) = 1 < n! = P(n,n)$ because there is just one (unordered) subset of all the n items in our pool, but $n!$ ways of putting that subset in order.

(e) By the counting formula, $C(n,n{-}k) = n! \,/\, (n{-}k!)(n{-}(n{-}k))! = n! \,/\, (n{-}k!)k! = n! \,/\, k!(n{-}k)! = C(n,k)$ again by the counting formula.

We note a trivial but very useful consequence of our formula for combinations. Suppose that we are given a set A with n elements, equipped with a *fixed order*. How many ways of choosing k items from A without repetition *in this fixed order*?

Reflection reveals that is the same as the number of ways of choosing k items from A *disregarding order* (and again without repetition), i.e. $C(n,k)$. The reason is that there is a bijection between the two classes of selections. Consider the function that takes an arbitrary subset X of A to the ordered tuple formed by writing its elements in the fixed order. Clearly this is injective and onto.

In general: *for counting selections from a set, fixed order on that set is equivalent to disregarding order on that set.* This simple (if rather roughly expressed) fact will help us establish our counting formula for the last mode of selection.

Alice Box: The principle of inclusion and exclusion

Hatter: This is a good moment for you to cash in the rain-check that you received in the first section of this chapter. Remember that we learned how to calculate the number of elements of $A{\cup}B$, for arbitrary finite sets A,B, given the number in each of A, B, and $A \cap B$.

Alice: Yes, $\#(A{\cup}B) = \#(A) + \#(B) - \#(A \cap B)$. That was the addition principle for two arbitrary sets, and I wanted to generalize it to n arbitrary sets.

Hatter: Let's begin by analysing what happened for two sets. We counted the elements of A and B separately and added them. But as the two sets may not

(Continued)

Alice Box: (Continued)

be disjoint, that counted the elements in $A \cap B$ twice, so we subtract them once.

Alice: And for three sets?

Hatter: For three sets the rule is:

$$\#(A \cup B \cup C) = \#(A) + \#(B) + \#(C) - \#(A \cap B) - \#(B \cap C) - \#(A \cap C) \\ + \#(A \cap B \cap C).$$

Here we begin by adding the number of elements in $A, B,$ and C. But some of these may be in two or more of the sets, so we will have *overcounted*, by counting those elements at least twice. So in the next stage we subtract the number of elements that are in the intersection of any two of the sets. But then we may have *undercounted*, for if there is an item x that is in all three sets, it was counted three times in the first stage and then subtracted three times in the second stage, and thus needs to be added in again, which is what we do in the last stage.

Alice: What's the general rule?

Hatter: Let $A_1,...,A_n$ be any sets with $n \geq 1$. The cardinality of their union $A_1 \cup ... \cup A_n$ is the sum of the following numbers:

+ (the sum of the cardinalities of the sets taken individually)

– (the sum of the cardinalities of the sets intersected two at a time)

+ (the sum of the cardinalities of the sets intersected three at a time)

– (the sum of the cardinalities of the sets intersected four at a time)

....

± (the sum of the cardinalities of the n sets intersected n at a time)

Alice: Why the ± at the end?

Hatter: Because it depends on whether n is even or odd: if n is odd, then it will be +, if n is even, then it is –. Of course, this rule can be written in a much more concise mathematical notation, but for our purposes, we can leave that aside.

Alice: How does that tie in with what we are doing now?

Hatter: Combinations! For example, in the second line of the rule, we need to work out the sum of the cardinalities of the sets $A_i \cap A_j$ for all distinct $i,j \leq n$. How many such sets are there? Well, there are $C(n,2)$ two-element subsets $\{A_i, A_j\}$ of the n-element set $\{A_1,...,A_n\}$, so there are $C(n,2)$ such intersections to consider. In other.

(*Continued*)

Alice Box: (Continued)

words, the second line of our rule asks us to make the sum of $C(n,2)$ numbers. Likewise the third line of the rule asks for the sum of $C(n,3)$ numbers, and so on

Alice: What is the name of this rule?

Hatter: Traditionally it is called the *principle of inclusion and exclusion*, because you first include too much then exclude too much. But personally I prefer to call it by an even more graphic name: the principle of *overcounting and undercounting.*

EXERCISE 5.4.6

(a) Draw a Euler diagram representing three intersecting sets, label the relevant areas, and paraphrase the rule of inclusion and exclusion in terms of the areas.

(b) Write out the principle of inclusion and exclusion in full for the case $n = 4$. Then calculate $C(4,k)$ for each $k = 1,..,4$ to check that at each stage you have made the right number of inclusions or exclusions.

5.5 Counting Formulae: Perms and Coms with Repetition

We now consider the two modes of selection that allow repetition of the selected elements. The table below adds these two modes into the table that we had for the non-repetitive selections.

Table 5.4 Formulae for four modes of selecting k items from an n-element set.

Mode of Selection	Notation	Standard Name	Formula	Proviso	Example: $n = 10$, $k = 6$
O+R−	$P(n,k)$	Permutations	$n! \,/\, (n{-}k)!$	$k \leq n$	151,200
O−R−	$C(n,k)$	Combinations	$n! \,/\, k!(n{-}k)!$	$k \leq n$	210
O+R+		Permutations with repetition	n^k	none	1,000,000
O−R+		Combinations with repetition	$(n{+}k{-}1)! \,/\, k!(n{-}1)!$	$n \geq 1$	5,005

5.5.1 The Formula for Permutations with Repetition Allowed (O+R+)

We are looking at the number of ways choosing of k elements from a n-element set, this time allowing repetitions and distinguishing orders. To fix ideas, keep in mind a simple example: how many six-digit telephone numbers can be concocted with the usual ten digits 0–9. The formula tells us that there are $10^6 = 1,000,000$. Far more than the figure of 151,200 for plain permutations where repetition is not allowed, and dwarfing the 210 for combinations. What is the reasoning?

As for plain permutations, we apply the multiplication principle. Clearly, there are n ways of choosing the first item. But, as repetition is allowed, there are again n ways of choosing the second, and so on, thus giving us $n \cdot \ldots \cdot n$ (k times) i.e. n^k possible selections.

In a more abstract language: each selection can be represented by an ordered k-tuple of elements of the basic set, so that the set of all the selections corresponds to the Cartesian product of the n-element set by itself k times, which we have seen by the principle of multiplication has cardinality n^k. As we remarked earlier, this is the same as the number of *functions* from a k-element set such as $\{1,\ldots,k\}$ into an n-element set such as $\{1,\ldots,n\}$.

Note that this operation makes perfectly good sense even when $k > n$. For example, if only two digits may be used, we can construct six-digit telephone numbers out of them, and there will be 2^6 of them.

EXERCISE 5.5.1 (WITH PARTIAL SOLUTION)

(a) What happens in the limiting case that $n = 1$? Does the figure provided by the formula square with intuition? Likewise for $n = 0$.

(b) What about the limiting case that $k = 1$? Does the figure given by the formula make intuitive sense? Similarly for $k = 0$.

(c) The drinks machine has three kinds of coffee, four kinds of soft drink, still water and sparkling water. Every working day I buy a coffee to wake up in the morning, a bottle of water to wash down lunch, and a soft drink for energy in the afternoon. Assuming that I start work on a Monday and work five days a week, how many weeks before I am obliged to repeat my daily selection?

Solution to (a), with comments: When $n = 1$ then $n^k = 1$ for any k. Intuitively, this is what we want, for there is only one way to write down the same thing k times. When $n = 0$ then $n^k = 0$ for any $k \geq 1$. This is also intuitively natural: in this case no selection can be made since there is nothing to select from. The ultra-limiting case where $n = 0$ and $k = 0$ is rather special: here $n^k = 0^0 = 1$, not 0.

One could accommodate this with intuition by reflecting that in this case, we are asked to select nothing from nothing, and by doing nothing we achieve just that, so that there is one selection (the empty selection) that does the job. This contrasts with the fact that if we are asked to select something from nothing (the case $n = 0$, $k \geq 1$), there is *no way* of doing it – not even by doing nothing! So intuition, carefully cultivated, accords with the formula after all.

That said, it should also be granted that, in mathematics, definitions sometimes give rather odd-looking outcomes in their limiting cases. Nevertheless, so long as the definition works as desired in the principal cases, the mathematician is pretty much free in the limiting ones to stipulate whatever permits the smoothest formulation and proof of general principles. Reflection on such limiting cases sometimes gives interesting philosophical perspectives.

5.5.2 The Formula for Combinations with Repetition Allowed (O–R+)

Example: A ten-person committee needs volunteers to handle six tasks. We are not interested in the order of the tasks, and dedicated members are allowed to volunteer for more than one task. How many different ways may volunteers come forward?

Since we are not interested in the order in which the tasks are performed, but we do allow multiple volunteering, we need to apply the formula for combinations with possible repetition in Table 5.4. This tells us that it is $(10+6-1)! \: / \: 6!(10-1)! = 15! \: / \: 6! \: 9! = 15 \cdot 14 \cdot 13 \cdot 12 \cdot 11 \cdot 10 \: / \: 6 \cdot 5 \cdot 4 \cdot 3 \cdot 2 = 7 \cdot 13 \cdot 11 \cdot 5 = 5{,}005$. More than the 210 for plain combinations, but still much less than the 151,200 for plain permutations, not to speak of the 1,000,000 for permutations with repetition allowed.

This mode was the trickiest to interpret in terms of functions in Section 5.3, and its counting formula is likewise the trickiest to prove. The basic idea of the argument is to re-conceptualize the question as one of selecting with neither order nor repetitions, i.e. using combinations, from a larger set of elements.

As mentioned in conversation with Alice, we are in effect counting the number of functions f from a set $\{a_1, \ldots, a_n\}$ or simply $\{1, \ldots, n\}$ with n elements into

$\{0,\ldots,k\}$ such that $f(a_1)+\ldots+f(a_n) = k$. Consider any such function f. To get the gestalt, we think of ourselves as provided with a sufficiently long sequence of slots, which we fill up to a certain point in the following way. We write 1 in each of the first $f(a_1)$ slots, then a label $+$ in the next slot, then 2 in each of the next $f(a_2)$ slots, then a label $+$ in the next one, and so on up to the $(n-1)$th label $+$, and finally write n in each of the next $f(a_n)$ slots, and stop. This makes sense provided $n \geq 1$. Careful: we do *not* put a $+$ at the end: we use $n-1$ plus-signs just as in the expression $f(a_1)+\ldots+f(a_n)$. The construction will thus require $k+(n-1)$ slots, since $f(a_1)+\ldots+f(a_n) = k$ and we insert $n-1$ plus-signs.

Now the slot pattern we have created is in fact fully determined by the total number $k+(n-1)$ of slots and the particular slots where the plus-signs go. This is because we can recuperate the whole pattern by writing 1 s in the empty slots to the left of the first plus-sign, then 2 s in the empty slots to the left of the second plus-sign, and so on up to $(n-1)$s in the empty slots to the left of the $(n-1)$th plus-sign and, not to be forgotten, ns in the remaining empty slots to the right of the last plus-sign until all $k+(n-1)$ slots are occupied. So in effect, we are looking at the number of ways of choosing $n-1$ slots for plus-signs from a set of $k+(n-1)$ slots. The $n-1$ plus-sign-slots must be distinct from each other (we can't put two plus-signs in the same slot), and we are holding the order of the items fixed.

Thus, recalling the observation made when discussing combinations – that for counting selections from a set, fixed order on that set is equivalent to disregarding order on that set – we have reduced applications of combination with repetition (O–R+) to rather larger applications of plain combination (O–R–). Specifically, we have shown that O–R+$(n,k) = C(k+n-1, n-1)$.

The rest is calculation: $C(k+n-1, n-1) = (k+n-1)! / (n-1)! ((k+n-1)-(n-1))!$ $= (k+n-1)! / (n-1)! k! = (n+k-1)! / k! (n-1)!$ – which is the counting formula in Table 5.4.

EXERCISE 5.5.2 (WITH PARTIAL SOLUTION)

(a) Use the formula to calculate the number of combinations with repetition of 6 items taken k at a time, for each $k = 1,\ldots,8$.

(b) The restaurant has five flavours of ice-cream, and you can ask for one ball, two balls, or three balls (unfortunately, not the same price). How many different servings are possible?

Solution to (b): We are looking at a set with 5 elements (the flavours), and at selections of one, two, or three items (the balls), allowing repetition. Does order count? Well, the formulation of the problem is rather vague. Two interpretations suggest themselves.

First interpretation. If the balls are laid out in a row on a suitably shaped dish, we might regard the serving strawberry/chocolate/vanilla as different from chocolate/vanilla/strawberry. In that case, we are in the mode O+R+, and reason as follows. Break the set of selections down into three disjoint subsets – those with one ball, two balls, three balls. Calculate each separately. There are $5^1 = 5$ selections with one ball, $5^2 = 25$ with two balls, $5^3 = 125$ with three balls. So there are in total $5+25+125 = 155$ possible selections.

Second interpretation. But if we don't care about the order in which the balls are distributed on the plate – for example, if they are set out in triangular fashion in a round dish, or haphazardly – then we are in mode O–R+, and reason as follows. Again, break the set of selections into three disjoint subsets, and calculate each separately. There are still 5 selections with one ball, since order becomes irrelevant in that case, and $(5+1-1)!$ / $1!(5-1)! = 5!/4! = 5$. But there are now only $(5+2-1)!$ / $2!(5-1)! = 6!$ / $2!4! = 3 \cdot 5 = 15$ selections with two balls, and just $(5+3-1)!$ / $3!(5-1)! = 7!$ / $3!4! = 7 \cdot 5 = 35$ with three balls, totalling only $5+15+35 = 55$ possible selections.

5.6 Rearrangements and Partitions

5.6.1 Rearrangements

How many ways of arranging the letters in *banana*? This sounds very simple, but its analysis is quite subtle. At first sight, it looks like a case of permutations allowing repetition: order definitely matters since *banana* $\neq$ *nabana*, and there are repeats since only three letters are filling six places. But there is also a big difference: *the amount of repetition is predetermined*: there must be exactly three *a*s, two *n*s, one *b*.

Selections of this kind are usually called *rearrangements*. We will follow this terminology, but you should be aware that the term is sometimes used more broadly for any transformation where order is significant.

It is important to begin with the right *gestalt*. In the case of '*banana*', don't think of yourself as selecting from a three-element set consisting of the letters a, n, b. Instead, focus on a six-element set consisting of the six *occurrences* of letters in the word, or six *slots* ready for you to insert those occurrences. This done, there are two ways of conceptualising this problem. They give the same numerical result, but provide an instructive contrast.

5.6.1.1 First Method

Let's look at the first approach. We think in terms of the six letter-occurrences. To distinguish them from letters-as-types we rewrite the word with subscripts: $b a_1 n_1 a_2 n_2 a_3$.

Analysis of the example: We know that there are $P(6,6) = 6! = 720$ permutations of the set $A = \{b_1, a_1, n_1, a_2, n_2, a_3\}$. Clearly dropping the subscripts from a identifies $3! = 6$ of these, and dropping the subscripts from n identifies another $2! = 2$ of them. Thus the number of rearrangements of the letters is $720 / 6 \cdot 2 = 60$.

Generalizing from 6 to n: Suppose we are given a word with n letter-occurrences, made up of k letters, with those letters occurring $m_1, \ldots, m_k$ times. Then there are $n! / m_1! \ldots m_k!$ rearrangements of the letters in the word.

Abstracting from words and letters: Given a set with A with n elements and a partition of A into k cells with $m_1, \ldots, m_k$ elements, the pointwise induced partition of the set of all permutations of A has $n! / m_1! \ldots m_k!$ elements.

By the *pointwise induced* partition we mean one that puts n-tuples $(a_1, \ldots, a_n)$ and $(a_1', \ldots, a_n')$ in the same cell iff a_i and a_i' are in the same cell as each other for all $i \leq n$.

So we are counting the cells of a specific partition – *not* of the given partition of A, but of the pointwise induced partition of the set of all permutations of A. This contrasts subtly with what will be done using the second method.

5.6.1.2 Second Method

The second approach begins on a rather more abstract level. This time we think in terms of six *slots* $s_1, \ldots, s_6$ ready to receive letter-occurrences.

Analysis of the example: The rearrangements of the letters can be thought of as functions f that take elements of the three-element set $\{a, n, b\}$ to *subsets* of $\{s_1, \ldots, s_6\}$ in such a way that $f(a)$ has three elements, $f(n)$ has two, $f(b)$ is a singleton, and the family $\{f(a), f(n), f(b)\}$ partitions $\{s_1, \ldots, s_6\}$. How can we count these functions?

We know that there are $C(6,3)$ ways of choosing a three-element subset for $f(a)$, leaving 3 slots unfilled. We have $C(3,2)$ ways of choosing a two-element subset of those three slots for $f(n)$, leaving 1 slot to be filled with the letter b. That gives us $C(6,3) \cdot C(3,2) \cdot C(1,1)$ selections. Calculating using the formula already known for combinations:

$$C(6,3) \cdot C(3,2) \cdot (1,1) = [6! / 3!(6-3)!] \cdot 3! / [2!(3-2)!] \cdot [1! / [1!(1-1)!]$$
$$= [6! / 3! \, 3!] \cdot [3! / 2!]$$
$$= 20 \cdot 3$$
$$= 60$$

which is the same figure as we obtained by the first method.

Generalizing from 6 to n: We have a set $\{a_1, \ldots, a_k\}$ of k letters to be written into a set $\{s_1, \ldots, s_n\}$ $(k \leq n)$ of slots, with each letter a_i being written in m_i slots. So we are looking at functions f that take elements of the set $\{a_1, \ldots, a_k\}$ to *subsets* of $\{s_1, \ldots, s_n\}$ in such a way that each of $f(a_i)$ has m_i elements and the family $\{f(a_i) : i \leq k)\}$ partitions $\{s_1, \ldots, s_n\}$. How can we count these functions?

We know that there are $C(n, m_1)$ ways of choosing an m_1-element subset for $f(a_1)$, leaving $n - m_1$ slots unfilled. We have $C(n - m_1, m_2)$ ways of choosing an m_2-element subset of those slots for $f(a_1)$, leaving $n - m_1 - m_2$ slots to be filled, and so on until all the n slots are used up. Clearly this can be done in the following number of ways:

$$C(n, m_1) \cdot C(n - m_1, m_2) \cdot \ldots \cdot C(n - m_1 - m_2 - \ldots - m_{k-1}, m_k)$$

This formula looks pretty ghastly. But if we write it out using the formula for combinations and then do successive cancellations, it simplifies beautifully, coming down to $n! \,/\, m_1! \ldots m_k!$ – which is the same formula as obtained by the other method.

Abstracting from words and letters: We want to count the number of functions f that take elements of a k-element set $\{a_1, \ldots, a_k\}$ to subsets of an n-element set $\{s_1, \ldots, s_n\}$ $(k \leq n)$ such that each set $f(a_i)$ has m_i elements and the family $\{f(a_i) : i \leq k)\}$ partitions $\{s_1, \ldots, s_n\}$. The formula is:

$$C(n, m_1) \cdot C(n - m_1, m_2) \cdot \ldots \cdot C(n - m_1 - m_2 - \ldots - m_{k-1}, m_k)$$
$$= n!/m_1! \ldots m_k!$$

EXERCISE 5.6.1

(a) Using the counting formula already known for combinations, write out the formula $C(n, m_1) \cdot C(n{-}m_1, m_2) \cdot \ldots \cdot C(n{-}m_1{-}m_2{-}\ldots{-}m_{k-1}, m_k)$ in full for the case that $n = 9$, $k = 4$. Perform the successive cancellations, and check that it agrees with $9! \,/\, m_1! \ldots m_4!$

(b) Apply the counting formula $n! \,/\, m_1! \ldots m_k!$ to determine how many ways of rearranging the letters in 'Turramurra'.

5.6.2 Counting Partitions with a Given Numerical Configuration

The second method suggests a further step that we can take. To explain this, first note that any partition of a finite set has a certain *numerical configuration*. This is a specification of the number k of cells of the partition, and of the number m_i $(1 \leq i \leq k)$

of elements in each of the cells. What the second method for counting rearrangements does, in effect, is count the number of *ordered partitions of A that are of a given numerical configuration.*

This immediately suggests a question: What is the number of plain (i.e. *unordered*) *partitions* of A with a given numerical configuration? The answer is almost as immediate. To forget the order of the k cells, simply divide by $k!$

Thus we have a *principle for counting partitions* with a given numerical configuration. The number of partitions of an n-element set A with a given numerical configuration (k cells, sizes $m_1,\ldots,m_k$) is given by the formula $n! \,/\, m_1!\ldots m_k!\,k!$

EXERCISE 5.6.2

There are 9 students in a class. How many ways of dividing them into tutorial groups of 2, 3 and 4 students for tutorials on Monday, Wednesday, and Friday?

Solution: We are looking at a set $A = \{a_1,\ldots,a_9\}$ with 9 elements, and we want to count ways of partitioning it into three cells of specified sizes. Do we regard the order of the cells as important? For example, do we wish to treat the division into cells C_1, C_2, C_3 as identical to the division into, say, cells C_2, C_1, C_3?

Suppose first that we *do not* wish to identify them. This would presumably be the case if we already intend to treat C_1 as the Monday tutorial, C_2 as the Wednesday one, and C_3 as the Friday one. Then we are looking at all the permutations of partitions with the specified cell sizes, and our counting formula should be $n! \,/\, m_1!\ldots m_k!$ giving us $9! \,/\, 2!3!4! = 1260$ possible ways of dividing up the students.

Suppose alternatively that we *do* wish to identify divisions that differ only in the order of their cells. This might be the case if for the present we are merely constituting the three groups, leaving open their scheduling. Then we are counting the number of partitions with the specified cell sizes, so our counting formula should be $n! \,/\, m_1!\ldots m_k!k!$, giving us only $9! \,/\, 2!3!4!3! = 1260\,/3! = 1260\,/6 = 210$ possible ways of dividing.

Comments on the solution: The answer we get depends on the method we use, which depends in turn on the exact interpretation of the problem. Sometimes there are only hints in the formulation to indicate which interpretation is meant. On other occasions the formulation may simply be ambiguous, in which case we say that the problem is *under-determined.*

Some texts take a perverse pleasure in keeping the hints to an almost imperceptible minimum in their problems. Not here.

We summarize the results of the section in the following table, which thus supplements Table 5.4 for the four O±R± modes.

Table 5.5 Formulae for counting partitions.

Description	Name	Counting Formula
Number of partitions with a given numerical configuration (k cells and cell sizes $m_1, \ldots, m_k$)	Partitions	$n! \,/\, m_1! \ldots m_k! \; k!$
Number of ordered partitions with a given numerical configuration (k cells and cell sizes $m_1, \ldots, m_k$)	Rearrangements, alias ordered partitions	$n! \,/\, m_1! \ldots m_k!$

Alice box: Different cultures, different languages

Alice: I have been reading some other textbooks on this material, and there is something that leaves me quite confused.

Hatter: Go ahead.

Alice: I find sentences that begin like this: 'Let A be a set of six elements, of which three are identical...'. But that doesn't make any sense!

Hatter: Why not?

Alice: You can't have different items that are identical. If $A = \{a, b, c, d, e, f\}$ and a, b, c are identical to each other, then A has at most four elements, not six. These books seem to be speaking a different language!

Hatter: Well, at least a different dialect, and it certainly can be confusing! Let me try to explain. The subject that we are calling 'sets, logic and finite mathematics' is in fact a fairly recent assemblage, made up of different topics that were developed at different times using specific terminologies and, most unsettlingly, conceptual structures that do not fit well with each other. Textbooks like ours try to put them all together in one coherent story, beginning with sets.

But some of the areas are much older than the theory of sets. For example, the theory of counting was already highly active in the seventeenth century, and indeed goes back many centuries before – consider Fibonacci in 1202! On the other hand, the theory of sets as we know it today, and its use to ground the

(Continued)

> *Alice Box:* (Continued)
>
> theories of relations, functions, partitions etc, appeared only in the late nineteenth century. As you would expect, the conceptual frameworks were quite different.
>
> Today, when authors bring various disciplines between the covers of a single book, they are torn between the demands of coherence of the integrated structure and faithfulness to the deeply rooted terminologies of specific topics. Usually something is given up on both sides. In this book, coherence of the broad picture tends to take priority.
>
> *Alice*: So how can I make sense of talk about selections from a set with several identical elements?
>
> *Hatter*: Above all, try to interpret what may be meant in the particular problem. Sometimes when it speaks of identical or indistinguishable elements, it means that we have no way of telling the difference between different orders of the elements. This is in effect a way of saying that we are interested in selections where order is immaterial but repetition is possible, i.e. O–R+.
>
> *Alice*: And in other cases?
>
> *Hatter*: Sometimes, reference to identical, indistinguishable or indiscernible elements is a way of pointing to a partition (in other words, an equivalence relation) between elements. Then we have a problem of counting partitions (or ordered partitions).
>
> *Alice*: How can I tell which is when?
>
> *Hatter*: No formal algorithm. It is an exercise in hermeneutics. Try to find the most natural interpretation of the problem, taking into account its content as well as its structure, and be attentive to little hints. There is not always a unique answer, as different readings may sometimes be plausible. The following exercise illustrates the task.

EXERCISE 5.6.3 (WITH HINTS FOR SOLUTION)

(a) How many ways are there to place 12 indistinguishable balls into 10 slots?

(b) There are 12 bottles of wine in a row, of which 5 are the same, another 4 the same, and another 3 the same. They are taken down and put into a box. Then they are put back on the shelf. How many different results are possible?

Hints for solution:

(a) All the balls are 'indistinguishable'. Presumably we are not thinking of the trivial partition with just one cell. We are envisaging the selection, with repetition permitted but order ignored, of 12 slots from the set of 10, i.e. 12-combinations with repetition from a 10-element set. Go on from there.

(b) From the phrasing, it appears that we are given a partition of the set of 12 bottles of wine according to their labels. The partition has three cells, and we are asked to count the rearrangements of the 12 items given that partition. Go on from there.

FURTHER EXERCISES

5.1. *Addition and multiplication principles and their joint use*

(a) A computer represents elements of $\mathbf{Z}$ using binary digits 0 and 1. The first digit represents the sign (negative or positive), and the remainder represent the magnitude. How many distinct integers can we represent with n binary digits? *Warning:* Be careful with zero.

(b) I want to take two books with me on a trip. I have two logic books, three mathematics books, and two novels. I want the two books that I take to be of different types. How many possible sets of two books can I take with me?

(c) In the USA, radio station identifiers consist of either 3 or 4 letters of the alphabet, with the first letter a K or a W. Determine the number of possible identifiers.

(d) You are in Paris, and you want to travel Paris-Biarritz-London-Paris or Paris-London-Biarritz-Paris. There are 5 flights each way per day between Paris and Biarritz, 13 each way between Paris and London, but only 3 per day from London to Biarritz and just 2 in the reverse direction. How many flight sequences are possible?

5.2. *The principle of inclusion and exclusion*

Amal telephones 5 different people, Bertrand telephones 10, Clarice telephones 7. But 3 of the people called by Amal are also called by Bertrand, 2 of those called by Bertrand are rung by Clarice, and 2 of those contacted by Clarice are called by Amal. One person was called by all three. How many people were called?

5.3. *Four basic modes of selection*

(a) In the first round of a quiz show, a contestant has to answer correctly seven questions each with a yes/no answer. Another contestant has to answer correctly two questions, each of which is multiple-choice with seven possible answers. Assuming that both contestants are totally ignorant and guess blindly, which faces the more daunting task?

(b) Australia has seven states. A manager proposes to test a product in four of those states. In how many ways may the test states be selected?

(c) Another manager, more cautious, proposes to test the product in one state at a time. In how many ways may a test-run be selected?

(d) Even more cautious, the director decides to test the product in one state at a time, with the rule that the test-run is terminated if a negative result is obtained at any stage. How many test-runs are possible?

(e) A game-board is a 4 by 4 grid of squares, and you have to move your piece from the bottom left square to the top right one. The only moves allowed are 'one square to the right' and 'one square up'. How many possible routes? *Warning*: This is more subtle than you may at first think. The answer is not 2^8.

(f) You have five dice of different colours, and throw one after another, keeping track of which is which. What is the natural concept of an outcome here? How many such outcomes are there?

(g) You throw five dice together, and are not able to distinguish one die from another. What is the natural concept of an outcome here, and how many are there?

5.4. *Rearrangements and partitions*

(a) The candidate wins a free holiday in the place whose name admits the greatest number of arrangements of its letters: 'Mississippi', 'Ouaga-doudou', 'Woolloomooloo'. Make the calculations and give the answer to win the trip.

(b) How many arrangements of the letters in 'Woolloomooloo' make all occurrences of any given letter contiguous?

(c) How many distributions of 10 problems are possible among Albert, Betty, Carlos and Deborah if Albert is to do 2, Betty 3, and Deborah 5. *Hint*: Conceptualize this as an ordered partition.

(d) How many ways of dividing 10 problems up into three groups of 2, 3 and 5 problems?

(e) Which is larger: the number of partitions of a set with 12 elements into three sets of four elements, or the number of partitions of the same set into four sets of three elements?

Selected Reading

Almost every text on discrete mathematics has a chapter on this material somewhere in the middle, and although terminology, notation and order of presentation differ from book to book the material covered is usually much the same. One clear beginner's exposition is:

Rod Haggarty *Discrete Mathematics for Computing*. Pearson, 2002, Chapter 6.

Other presentations include:

John Dossey et al. *Discrete Mathematics*. Pearson, 2006 (fifth edition), Chapter 8.

James L. Hein *Discrete Structures, Logic and Computability*. Jones and Bartlett, 2005 (second edition), Chapter 5.3.

Richard Johnsonbaugh *Discrete Mathematics*. Pearson, 2005 (sixth edition), Chapter 6.

Seymour Lipschutz *Discrete Mathematics*. McGraw Hill Schaum's Outline Series, 1997, Chapter 6.

6
Weighing the Odds: Probability

Chapter Outline

In this chapter, we introduce the elements of probability theory. In the spirit of the book, we confine ourselves to the discrete case, i.e. probabilities on finite domains, leaving aside the infinite one.

We begin by defining *probability functions* on a finite sample space and identifying some of their basic properties. So much is simple mathematics. This is followed by some words on different philosophies of probability, and warnings of traps that arise in applications. Then back to the mathematical work, introducing the concept of *conditional probability* and setting out its properties, its connections with *independence*, and its role in *Bayes' theorem*. In the final section we explain the notions of a '*random variable*' or payoff function, *expected value*, and *induced probability* distribution.

6.1 Finite Probability Spaces

In the chapter on principles of counting (alias combinatorics), we remarked that the area is rather older than the modern theory of sets, and tends to keep some of its traditional ways of speaking even when the ideas can be expressed in a standard set-theoretic manner. The same is true of probability theory, creating problems for both reader and writer of a textbook like this. If we simply follow the rather loose traditional language, the reader may fail to see how it rests on fundamental notions. On the other hand, if we translate everything into set-theoretic terms the reader is ill-prepared for going on to other expositions.

For this reason we follow a compromise approach. We make use of traditional probabilistic terminology, but at each step show how it is really serving as short-hand for a uniform one in terms of sets and functions. A table will be used to keep a running tally of the correspondences.

It turns out that applications of probability theory to practical problems often need to deploy the basic counting principles that were developed in the preceding chapter, and so the reader should be ready to flip back and forth to refresh the mind whenever needed.

6.1.1 Basic Definitions

The first concept that we need in discrete probability theory is that of a *sample space*. Mathematically, this is just an arbitrary finite (but non-empty) set S. The term 'sample space' merely indicates that we intend to use it in a probabilistic framework.

A *probability distribution* (or just *distribution* for short) is an arbitrary function $p: S \to [0,1]$ such that $\Sigma\{p(s): s \in S\} = 1$, i.e. the sum of the values $p(s)$ for $s \in S$ is equal to one. Recall that $[0,1]$, often called *the real interval*, is the set of all real numbers from 0 to 1 included, i.e. $[0,1] = \{x \in \mathbf{R}: 0 \le x \le 1\}$. Actually, in the context of discrete probability, i.e. where S is a finite set, we could without loss of generality reduce the target of the probability function to the set of all rational numbers from 0 to 1. But we may as well allow the whole of the real interval, which gives us no extra trouble and will be needed if ever you pass to the infinite case.

EXERCISE 6.1.1 (WITH SOLUTION)

Let S be the set $\{a,b,c,d,e\}$. Which of the following functions p are probability distributions on S? Give a reason in each case.

(a) $p(a) = 0.1$, $p(b) = 0.2$, $p(c) = 0.3$, $p(d) = 0.4$, $p(e) = 0.5$.

(b) $p(a) = 0.1$, $p(b) = 0.2$, $p(c) = 0.3$, $p(d) = 0.4$, $p(e) = 0$.

(c) $p(a) = 0$, $p(b) = 0$, $p(c) = 0$, $p(d) = 0$, $p(e) = 1$.

(d) $p(a) = -1$, $p(b) = 0$, $p(c) = 0$, $p(d) = 1$, $p(e) = 1$.

(e) $p(a) = p(b) = p(c) = p(d) = p(e) = 0.2$.

Solution: (a) No: the values do not add to one. (b) Yes: values are in the real interval and add to one. (c) Yes: same reason. (d) No: not all values are in the real interval. (e) Yes: values are in the real interval and add to one.

The last of the functions in the exercise is clearly a very special one: all the elements of the sample space receive the same value. This is known as an *equiprobable* (or *uniform*) *distribution*, and is particularly easy to work with. Given a sample space S with n elements, there is evidently just one equiprobable distribution $p: S{\rightarrow}[0,1]$, and it puts $p(s) = 1/n$ for all $s \in S$. But it should be understood that this is a special case, not only mathematically but also in practice. Many problems involve unequal distributions, and we cannot confine ourselves to equiprobable ones.

Given a distribution $p: S{\rightarrow}[0,1]$, we can extend it to a function p^+ on the power set $\mathcal{P}(S)$ of S by putting $p^+(A) = \Sigma\{p(s): s \in A\}$ when A is any non-empty subset of S, and $p^+(A) = 0$ in the limiting case that $A = \emptyset$. This is the *probability function* determined by the distribution p and, to keep notation down, we usually drop the superscript and also call it p. Note that even for a very small set S there will be infinitely many distributions on S, but each of them determines a unique probability function on $\mathcal{P}(S)$ or, as one also says, *over S*. The pair made up of the sample space S and the probability function $p: \mathcal{P}(S){\rightarrow} [0,1]$ is called a *probability space*. Traditionally, the elements of $\mathcal{P}(S)$, i.e. the subsets of S, are called *events*.

6.1.2 Properties of Probability Functions

Some basic properties of probability functions may be obtained easily from the definition.

EXERCISE 6.1.2 (WITH SOLUTION)

Let $p: \mathcal{P}(S){\rightarrow} [0,1]$ be a probability function over the sample space S. Show that p has the following properties (where A, B, are arbitrary subsets of S).

(a) $p(\emptyset) = 0$.

(b) $p(S) = 1$.

(c) $p(A{\cup}B) = p(A) + p(B) - p(A{\cap}B)$.

(d) When $A{\cap}B = \emptyset$ then $p(A{\cup}B) = p(A) + p(B)$.

(e) $p(S \backslash A) = 1 - p(A)$.

(f) $p(A \cap B) = p(A) + p(B) - p(A \cup B)$.

(g) When $A \cup B = S$ then $p(A \cap B) = p(A) + p(B) - 1 = 1 - [p(S \backslash A) + p(S \backslash B)]$.

Solution:

(a) Explicit in the definition of a probability function, limiting case.

(b) $p(S) = \Sigma\{p(s): s \in S\} = 1$ by the definition of a probability function (for the first equality) and the definition of a distribution (for the second).

(c) $p(A \cup B) = \Sigma\{p(s): s \in A \cup B\} = \Sigma\{p(s): s \in A\} + \Sigma\{p(s): s \in B\} - \Sigma\{p(s): s \in A \cap B\} = p(A) + p(B) - p(A \cap B)$.

(d) By (c) noting that when $A \cap B = \varnothing$ then $p(A \cap B) = p(\varnothing) = 0$.

(e) $S = A \cup (S \backslash A)$ and the sets A, $S \backslash A$ are disjoint, so by (d) we have $p(S) = p(A) + p(S \backslash A)$ and thus by arithmetic $p(S \backslash A) = p(S) - p(A) = 1 - p(A)$ by (b).

(f) This can be shown from first principles, but it is easier to get it by arithmetic manipulation of (c).

(g) The first equality is immediate from (f) noting that when $A \cup B = S$ then $p(A \cup B) = p(S) = 1$ by (b). For the second equality, $p(A) + p(B) - 1 = (1 - p(S \backslash A)) + (1 - p(S \backslash B)) - 1$ by (e), which then simplifies arithmetically $1 - [p(S \backslash A) + p(S \backslash B)]$.

Comment: Note how once we have (a–c), the others follow without going back to the definitions of distribution or probability function. Conversely, it is possible to get (c) by adroit exploitation of its special case (d).

In the above exercise, we used the notation $S \backslash A$ for the complement of A with respect to the sample space S, in order to avoid any possible confusion with the arithmetic operation of subtraction. Although they are closely related, they are of course different. From now on, however, we will write more briefly $-A$.

It can be quite tiresome to write out the function sign p over and over again when calculating or proving in probability theory. For problems using only one probability function we can use a more succinct notation. For each subset A of the sample space, we can write $p(A)$ as $\underline{A}$ with the underlining representing the probability function p. For example, we may write the equation $p(A \cup B) = p(A) + p(B) - p(A \cap B)$ as $\underline{A \cup B} = \underline{A} + \underline{B} - \underline{A \cap B}$.

But be careful! If the problem involves more than one probability function (as for example in the iterated conditionalizations later in this chapter) this shorthand notation cannot be used, since it cannot keep track of the different functions. For this reason, your instructor may not like you to use it at all, so check on class policy before using it anywhere outside rough work. In what follows, we will stick to standard presentation.

EXERCISE 6.1.3

(a) Use the results of Exercise 6.1.2 to show that when $A \subseteq B$ then $p(A) \leq p(B)$.

(b) Use this to show that $p(A \cap B) \leq p(A)$ and also $p(A \cap B) \leq p(B)$.

(c) Let $S = \{a,b,c,d\}$ and let p be the equiprobability distribution on S. Give examples of events, i.e. sets $A,B \subseteq S$, such that respectively $p(A \cap B) < p(A) \cdot p(B)$, $p(A \cap B) = p(A) \cdot p(B)$, $p(A \cap B) > p(A) \cdot p(B)$.

The following table keeps track of translations between traditional probabilistic and modern set-theoretic terminology. The first seven rows should be clear from what we have done so far; the other rows will be explained as we continue in the chapter.

Table 6.1 Two languages for probability.

Traditional probabilistic	Modern set-theoretic
sample space S	arbitrary set S
sample point	element of S
probability distribution	function p: $S \to [0,1]$ whose values sum to 1
event	subset of S
elementary event	singleton subset of S
null event	empty set
mutually exclusive events	disjoint subsets of S
experiment	situation to be modelled probabilistically
$p(A|B)$	$p(A,B)$
random variable X	function f: $S \to \mathbf{R}$
range space R_X	range $f(S)$ of a function f: $S \to \mathbf{R}$
distribution of X	$p \circ f^{-1}$ for function f: $S \to \mathbf{R}$ and distribution p: $S \to [0,1]$
$P(X = x)$	$p \circ f^{-1}(x) = p(\{s \in S: f(s) = x\})$ for $x \in \mathbf{R}$

6.2 Philosophy and Applications

Probability theory differs from any of the topics that we have studied so far, in two respects, one philosophical and the other practical.

Philosophically, the intuitive notion of probability is much more slippery than that of a set. It is more perplexing than the notion of truth, which we used in logic boxes to define connectives such as negation, conjunction, disjunction and material implication. When we declare that an event is, say, highly improbable we are not committing ourselves to saying that it does not take place. Indeed, we may have a collection of events, each of which known to be highly improbable while we also know that at least one *must* take place. Consider for example a properly conducted lottery with many tickets. For each individual ticket, it is highly improbable that it will win; but it is nevertheless certain that one will do so.

So what is meant by saying that an event is probable, or improbable? There are two main kinds of perspective on the question.

- One of them sees probability as a measure of uncertainty, understood as a state of mind. This is known as the *subjectivist* conception of probability. The general idea is that a probability function p is a measure of the degree of belief that a fully rational (but not omniscient) agent would have concerning various possible events, given some limited background knowledge that is not in general enough to believe or disbelieve them without reservation.

- The other sees probability as a manifestation of as something in the world itself, independent of the thinking of any agent. This is known as the *objectivist* conception of probability. Often, in this approach, the probability of an event is taken to be a measure of the long-run tendency for events similar to it to take place in contexts similar to the one within under consideration. This particular version of the objectivist approach called the *frequentist* conception.

Of course, these thumbnail sketches open more questions than they begin to answer. For example, under the subjectivist approach, what counts is not your degree of belief or mine, nor even that of some reputed expert or guru, but rather the degree that a *rational agent* would have. So how are we are to understand rationality in this context? Surely, in the final analysis, rationality is not such a subjective matter. Properly unpacked, the subjectivist approach may thus end up being not so subjective after all!

On the other hand, the frequentist view leaves us with the difficult problem of explaining what is meant by 'contexts similar to the one within under consideration'. It also leaves us with the question of why the long-run frequency of something should give us any confidence at all in the short term. As the saying goes, in the long run we are all dead. It is difficult to justify confidence in real-life decisions, which need to be taken within a limited time, in terms of the frequentist approach.

It is often felt by practitioners that some kinds of situation are more easily conceptualized in subjectivist terms, others in frequentist ones. The subjectivist approach seems to fit better for events that are rare or unique, or about which little data on past occurrences of similar events is available. For example, *prima facie* it may be difficult for a frequentist to deal with the question of the probability of the earth being rendered uninhabitable by a nuclear war in the coming century. On the other hand, there are other kinds of question on which lots of data are available, on which frequencies may be calculated, and for which the frequentist conception of probability seems appropriate. That is how insurance companies keep afloat.

As this is not a text of philosophy, we will not take such issues any further. We simply note that they are difficult to formulate meaningfully, tricky to handle, and have not obtained any consensual resolution despite the centuries in which probability theory has been studied mathematically and applied to everyday affairs. The amazing thing is that we are able to carry on with both the mathematics of probability theory and its application to the real world, without resolving these philosophical issues. But this application is an art as much as a science, and more subtle than the mathematics itself.

The first point to realize is that you cannot get probabilities out of nothing. You have to make assumptions about the probabilities of some events in order to deduce probabilities about others. Such assumptions should always be made explicit, so that they can be recognized for what they are and assessed. Whereas in textbook examples, the assumptions are often simply given as part of the problem statement, in real life they can be difficult or impossible to determine or justify.

In practice, this means that whenever confronted with a problem of probabilities, two steps must be made before attempting any calculation.

- *Identify the sample space of the problem.* What is the set S that we take as domain of the probability distribution in the question?

- *Specify the values of the distribution.* Can it reasonably be assumed to be a uniform distribution? If so, we already have values for the function p at each point in the sample space (provided we know the size of the space), and we are ready to calculate. If not, we must ask: Is there any other distribution that is reasonable to assume?

Many students are prone to premature calculation, and training is needed to overcome it. In this chapter, we will always be very explicit about the sample space and the values of the distribution.

Alice Box: Probability in the infinite case

Alice: I know that we are studying only the finite case, but could you tell me just a little about what happens in the infinite one? Is it very different?

Hatter: When the sample space S is infinite, the question arises what to do with pairwise disjoint infinite unions. It is not always reasonable to extend the principle of addition to cover them. For example, if we look at the problem of choosing an arbitrary point from a square, it is reasonable to regard each point as equiprobable with the others. Thus, either all points in the square get zero probability or they all get the same non-zero probability r. But in neither case do these probabilities sum to one. In the former case they sum to zero. In the latter case they cannot sum to one; indeed, they cannot have a finitely large sum. This follows from the *principle of Archimedes*: for every real $r > 0$ there is an integer n with $n \cdot r > 1$. From this it follows immediately that for every real $r > 0$ and every integer k there is an integer m with $m \cdot r > k$ (just put $m = kn$).

Alice: So what do we do?

Hatter: There are essentially two paths open. One, rather radical, is to widen the target set to contain not only real numbers but also 'infinitesimals' as defined and exploited in what is called non-standard analysis. The other, more conservative and usually followed, is to stick with the ordinary reals but redefine the notion of a probability function. Instead of taking its domain to be all of $\mathcal{P}(S)$, we take it to be *any field of subsets* of S, i.e. any non-empty collection of subsets of S that is closed under the operations of taking complements and *finite* unions (and thus also finite intersections). When S is finite, every field of subsets of S is the power set of some subset $S' \subseteq S$; but when S is infinite we can have many others. For example, the collection of all subsets of S that are either finite or cofinite (i.e. have finite complements with respect to S) is a field of sets.

Alice: And the principle of addition in this framework?

Hatter: Once again there are two main policies. The more conservative one is to rest content with finite additions only. The more adventurous is to require that the field of subsets is also closed under *countable* pairwise disjoint unions and intersections and, if the field is more than countable, assume a principle of addition for the probability of the union of a countable family of pairwise disjoint sets. But this is going way beyond discrete mathematics.

6.3 Some Simple Problems

In this section we run through a series of problems, to give practice in the task of applying the first principles of probability theory. The examples become progressively more sophisticated as they bring in additional devices.

Example 1. Throw a fair die. What is the probability that the number (of dots on the uppermost surface) is divisible by 3?

Solution and Comments: First we specify the sample space. In this case it is the set $S = \{1,2,3,4,5,6\}$. Next we specify the distribution. In this context, the term *fair* means that the questioner is assuming that we have an equiprobable distribution. The distribution thus puts $p(n) = 1/6$ for all positive $n \leq 6$. The event we have to consider is $A = \{3,6\}$ and so $p(A) = 1/6 + 1/6 = 1/3$.

Example 2. Select a playing card at random from a standard European pack. What is the probability that it is a hearts? That it is a court card (jack, queen, king)? That it is both? That it is hearts but not a court card? That it is either?

Solution and Comments: First we specify the sample space. In this case it is the set S of 52 cards in the standard European pack. Then we specify the distribution. In this context, the term *random* also means that we are assuming that it is an equiprobable distribution. The distribution thus puts $p(c) = 1/52$ for all cards c. We have five events to consider. The first two we call H, C and the others are $H \cap C$, $H \setminus C$, $H \cup C$. In a standard pack there are 13 hearts and 12 court cards, so $p(H) = 13/52 = 1/4$ and $p(C) = 12/52 = 3/13$. There are only 3 cards in $H \cap C$ so $p(H \cap C) = 3/52$, so there are $13-3 = 10$ in $H \setminus C$ giving us $p(H \setminus C) = 10/52 = 5/26$. Finally, $\#(H \cup C) = \#(H) + \#(C) - \#(H \cap C) = 13 + 12 - 3 = 22$ so $p(H \cup C) = 22/52 = 11/26$.

Example 3. An unscrupulous gambler has a loaded die, with probability distribution $p(1) = 0.1$, $p(2) = 0.2$, $p(3) = 0.1$, $p(4) = 0.2$, $p(5) = 0.1$, $p(6) = 0.3$. Which is more probable, that the die falls with an even number, or that it falls with a number greater than 3?

Solution and Comments: As in the first problem, the sample space is $S = \{1,2,3,4,5,6\}$. But this time the distribution is not equiprobable. Writing E and G for the two events in question, we have $E = \{2,4,6\}$ while $G = \{4,5,6\}$, so $p(E) = p(2) + p(4) + p(6) = 0.2 + 0.2 + 0.3 = 0.7$, while $p(G) = p(4) + p(5) + p(6) = 0.2 + 0.1 + 0.3 = 0.6$. Thus it is slightly more probable (difference $1/10$) that this die falls even than that it falls greater than 3.

Example 4. We toss a fair coin three times. What is the probability that at least two consecutive heads appear? The probability that exactly two heads (not necessarily consecutive) appear?

Solution and Comments: The essential first step is to get clear about the sample space. It is not the pair {heads, tails} or {H,T} for short. It is the set of all ordered triples with elements from {H,T}, i.e. {H,T}^{3}. Enumerating it in full, $S=$ {HHH, HHT, HTH, HTT, THH, THT, TTH, TTT}. As the coin is supposed to be fair, these are considered equiprobable. Let A be the event of getting at least two consecutive heads, and B that of getting exactly two heads (in any order). Then $A=$ {HHH, HHT, THH} while $B=$ {HHT, HTH, THH}. Thus $p(A) = 3/8 = p(B)$.

Alice Box: Repeated tosses

Alice: Not so fast! Something is funny here.

Hatter: What's wrong?

Alice: You seem to be making a hidden assumption. We are told that the coin is fair, so $p(H) = 0.5 = p(T)$. But how do you know that, say, $p(HHH) = p(HTH) = 1/8$? Perhaps the fact of landing heads on the first toss makes it less likely to land heads on the second toss. In that case the distribution over this sample space will not be equiprobable.

Hatter: Nice point! In fact, we *are* implicitly relying on such an assumption. We are assuming that $p(H)$ *always* has the value 0.5 no matter what happened on earlier tosses. Another way of putting it: the successive tosses have no influence on each other – H_{i+1} is independent of H_i, where H_i is the event 'heads on the ith toss'. We will be saying more about independence later in the chapter.

Example 5. Seven candidates are to be interviewed for a job. Of these, four are men and three are women. The interviews are conducted in random order. What is the probability that all the women are interviewed before any of the men? *Solution and Comments:* Let $M = \{m_1,\ldots,m_4\}$ be the four male candidates, and $W = \{w_1,\ldots,w_3\}$ be the three female ones. Then $M\cup W$ is the set of all candidates. We take the sample space to be the set of all permutations (i.e. selections in mode O+R−) of this seven-element set. There are $P(7,7) = 7!$ such permutations, and we assume that they are equiprobable. An interview in which all the women come before the men will be one in which we have some permutation of the women, followed by some permutation of the men, and there are clearly $P(3,3)\cdot P(4,4)$ of these. So the probability of such an interview pattern is $P(3,3)\cdot P(4,4)/P(7,7) = 3!4!/7! = 1/35$.

This is the first example making use of a concept from the chapter on counting. The sample space consisted of all permutations of a certain set and, since the space is assumed equiprobable, we are calculating the proportion of them with a certain property. If on the other hand, a problem requires a sample space consisting of combinations, we need to apply the appropriate formula for counting them, as in the following example.

Example 6. Of the 10 envelopes in my letterbox, 2 contain bills. If I pick 3 envelopes at random to open today, leaving the others for tomorrow, what is the probability that none of these three is a bill.

Solution and Comments: We are interested in the proportion of 3-element subsets of S that contain no bills. So we take the sample space S to consist of the set of all 3-element subsets of a 10-element set. Thus S has $C(10,3) = 10!/3!7!= 120$ elements. How many of these subsets have no bills? There are 8 envelopes without bills, and the number of 3-element subsets of this 8-element set is $C(8,3) = 8!/3!5!= 56$. So the probability that none of our three selected envelopes contains a bill is $56/120 = 7/15$ – just under 0.5, unfortunately.

Alice Box: Combinations or permutations?

Alice: Can't we do it with permutations too?

Hatter: Let's try. What's your sample space?

Alice: The set of all 3-element permutations of a 10-element set. This has $P(10,3) = 10!/7!= 10 \cdot 9 \cdot 8 = 720$ elements. And we are interested in the 3-element permutations of the 8-element no-bill set, of which there are $P(8,3) = 8!/5!= 8 \cdot 7 \cdot 6 = 336$ elements. As the sample space is equiprobable, the desired probability is given by the ratio $336/720 = 7/15$, the same as by the other method.

Hatter: So this problem may be solved by either method. However, the one using combinations has a smaller sample space. In effect, with permutations we are processing redundant information (the different orderings), giving us much larger numbers in the numerator and denominator, which we then cancel down. So it involves rather more calculation. Of course, much of the calculation could be avoided by cancelling as soon as possible, before finding fully numerical values for numerator and denominator. But as a general rule it is better to keep the sample space down from the beginning.

EXERCISE 6.3.1

(a) A pair of fair dice is thrown. What is the probability that the sum of the dots is 8? What is the probability that it is at least 5?

(b) Your debit card has a 4-digit pin number. You lose the card, which is found by a dishonest person. The cash dispenser allows 3 attempts to enter the pin number. What is the probability that the person accesses your account, assuming that he/she enters candidate pin numbers at random, without repetition (but of course allowing repetition of individual digits).

(c) There are 25 people in a room. What is the probability that at least two of the people are born on the same day of the year (but not necessarily the same year)? For simplicity, assume that each year has 365 days, and that the people's birthdays are randomly distributed through the year (both assumptions being, of course, rough approximations). *Hints*: (1) Be careful with your choice of sample space. (2) First find the probability that no two distinct people have the same birthday. (3) You will need a pocket calculator.

6.4 Conditional Probability

You throw a pair of fair dice. You already know how to calculate the probability of the event A that at least one of the dice is a 3. Your sample space is the set of all ordered pairs (n,m) where $n,m \in \{1,\ldots,6\}$ and you assume that the probability distribution on this space is equiprobable. There are 11 pairs (n,m) with either $n = 3$ or $m = 3$ (or both) so, by the assumption of equiprobability, $p(A) = 11/36$.

But what if we are informed, before looking at the result of the throw, that the sum of the two dice is 5? What is the probability that at least one die is a 3, *given that* the sum of the two dice is 5? This is known as *conditional probability*.

In effect, we are changing the sample space – restricting it from the set S of all 36 ordered pairs (n,m) where $n,m \in \{1,\ldots,6\}$, to the subset B consisting of those ordered pairs whose sum is 5. But it is very inconvenient to have to change the sample space each time we calculate a conditional probability. It is better to keep the sample space fixed and obtain the same result in another, equivalent, way. So we define the probability of A (e.g. that at least one die is a 3) *given B* (e.g. that the sum of the two faces is 5) as the ratio of $p(A \cap B)$ to $p(B)$. Writing the conditional probability of A given B as $p(A|B)$, we put:

$$p(A|B) = p(A \cap B)/p(B)$$

In the example given, there are four ordered pairs whose sum is 5, namely the pairs (1,4), (2,3), (3,2), (4,1), so $p(B) = 4/36$. Just two of these contain a 3, namely the pairs (2,3), (3,2), so $p(A \cap B) = 2/36$. Thus $p(A|B) = (2/36)/(4/36) = 2/4 = 1/2$ – larger than the unconditional (alias *prior*) probability $p(A) = 11/36$.

Note that the conditional probability of A given B is not in general the same as the conditional probability of B given A. While $p(A|B) = p(A \cap B)/p(B)$, we have $p(B|A) = p(B \cap A)/p(A)$. The numerators are the same, since $A \cap B = B \cap A$, but the denominators are different: they are $p(B)$ and $p(A)$ respectively.

EXERCISE 6.4.1

Calculate the value of $p(B|A)$ in the die-throwing example above, to determine its relation to $p(A|B)$.

The non-convertibility of conditional probability should not be surprising. In logic we know that the proposition $X \rightarrow Y$ is not the same as $Y \rightarrow X$, and in set theory the inclusion $X \subseteq Y$ is not equivalent to $Y \subseteq X$. All three are manifestations of the same failure of commutation or conversion.

Alice Box: Division by zero

Alice: One moment! There is something that worries me in the definition of conditional probability. What happens to $p(A|B)$ when $p(B) = 0$? This will be the situation, for example, when B is empty set. In this case we can't put $p(A|B) = p(A \cap B)/p(B) = p(A \cap B)/0$, since division by zero is not possible.

Hatter: Indeed, you are right! Division by zero is undefined in arithmetic, and so the definition that we gave for conditional probability does not make sense in this limiting case.

Alice: So what should we do?

Hatter: There are two main reactions. The standard one, which we will follow, is to treat conditional probability, like division itself, as a *partial function* of two arguments. Its domain is not $\mathcal{P}(S)^2$ where S is the sample space, but rather $\mathcal{P}(S) \times \{X \subseteq S: p(X) \neq 0\}$. The second argument ranges over all the subsets of the sample space S whose probability under the function p is greater than zero. That excludes the empty set $\varnothing$, and it may also exclude some other subsets since it can happen that $p(X) = 0$ for some non-empty $X \subseteq S$.

Alice: And the other reaction?

Hatter: Reconstruct probability theory, reversing the roles of conditional and unconditional probability. In other words, begin by taking a probability function on the sample space S to be any two-place function $p: \mathcal{P}(S)^2 \rightarrow [0,1]$ on the entire Cartesian product that satisfies certain conditions, and then introduce unconditional probability as a limiting case, putting $p(A) = p(A,S)$. This

(Continued)

Alice Box: (Continued)

approach is popular among philosophers, even though the conditions required of the two-place functions are rather less intuitive than those of the standard approach and different versions give distinct treatments of $p(A,B)$ when $p(B)$ = $p(B,S)$ = 0. In general, however, mathematicians and statisticians are quite happy with the standard approach that treats conditional probability as a partial function.

Taking into account Alice's question and the Hatter's explanation, we give the *rigorous definition of conditional probability*. Let $p: \mathcal{P}(S) \to [0,1]$ be any probability function on a (finite) sample space S. Then the corresponding conditional probability function on $\mathcal{P}(S) \times \{X \subseteq S: p(X) \neq 0\}$ is defined by the rule:

$p(A|B) = p(A \cap B)/p(B)$ in the (principal) case that $p(B) \neq 0$

$p(A|B)$ is undefined in the (limiting case) that $p(B) = 0$.

EXERCISE 6.4.2

(a) Suppose that we know from records that, in a certain population, the probability $p(C)$ of high cholesterol is 0.3 and the probability $p(C \cap H)$ of high cholesterol and heart attack together is 0.1. Find the probability $p(H|C)$ of heart attack *given* high cholesterol.

(b) Suppose that we know from records from another population that the probability $p(H)$ of heart attack is 0.2 and the probability $p(H \cap C)$ of heart attack and high cholesterol together is 0.1. Find the probability $p(C|H)$ of high cholesterol *given* heart attack.

In the empirical sciences, conditional probability is often employed in the investigation of causal relationships. For example, we may be interested in the extent to which high cholesterol leads to heart attack, and use conditional probabilities like those in the exercise above when analysing the data. But care is needed. We are entitled to look at the probability of cholesterol (suspected cause) given heart attack (undesirable event) just as much as at the converse probability of heart attack (suspected effect) given cholesterol (preceding condition). As far as the mathematical theory of probability is concerned, in a conditional probability $p(A|B)$ there is no need for B to be a cause of, or even temporally prior to, A. Indeed, the mathematics says nothing whatsoever about any kind of causal or temporal relationship between A and B in either direction.

For example, in Exercise 6.4.1, we calculated and compared the conditional probabilities $p(A|B)$ and $p(B|A)$, where A is the event that at least one die is a 3 and B is the event that the sum of the two faces is 5. Neither of these events can be thought of as causally related to the other, nor does either happen before or after the other. These conditional probabilities are merely a matter of numbers.

Even when A is an event like a heart attack and B one like high cholesterol, a high value of $p(A|B)$ does not imply that B is the cause (or even one among various contributing causal factors) of A. Reasoning from high conditional probability to causation is fraught with philosophical difficulties and hedged with practical provisos. Even reasoning in the reverse direction is not as straightforward as may appear. We will not attempt to unravel these issues, remaining within the mathematics of probability itself.

EXERCISE 6.4.3 (WITH SOLUTION)

(a) In a writers' association, 60% of members write novels, 30% write poetry, but only 10% write both novels and poetry. What is the probability that an arbitrarily chosen member writing poetry also writes novels? And the converse conditional probability?

(b) In the same writers' association, 15% write songs, all of whom also write poetry, and 80% of the songsters write love poetry. But none of the novelists write love poetry. What is the probability that (i) an arbitrary poet writes songs, (ii) an arbitrary songster writes poetry, (iii) an arbitrary novelist writes love poetry, (iv) an arbitrary novelist writing love poetry also writes songs, (v) an arbitrary songster writes novels?

Solution: We use the obvious acronyms N, P, S, L, for those who write novels, poetry, songs, and about love.

(a) $p(N|P) = p(N \cap P)/p(P) = 0.1/0.3 = 1/3$; while $p(P|N) = p(P \cap N)/p(N) = 0.1/0.6 = 1/6$.

(b) (i) $p(S|P) = p(S \cap P)/p(P)$. But by hypothesis, $S \subseteq P$, i.e. $S \cap P = S$, so $p(S|P) = p(S)/p(P) = 0.15/0.3 = 0.5$.

(ii) $p(P|S) = p(P \cap S)/p(S) = p(S)/p(S) = 1$.

(iii) $p(L \cap P | N) = p(L \cap P \cap N)/p(N) = p(\emptyset)/p(N) = 0/p(N) = 0.$

(iv) $p(S | L \cap P \cap N)$ is not defined since $p(L \cap P \cap N) = 0.$

(v) $p(N | S) = p(N \cap S)/p(S).$

Unfortunately, the data given in the problem do not suffice to determine the value of $p(N \cap S)$, so we cannot solve the problem fully. However, the data do give us an upper bound on the value of $p(N \cap S)$. Since 15% of the association are songsters and 80% of these write love poetry, we know that at least 12% of the association are songsters writing love poetry. Since none of the love poets are novelists, at least 12% of the association are songster non-novelists. But only 15% of the association are songsters in the first place, so at most 3% of the songsters are novelists, i.e. $p(N \cap S) \leq 0.03.$ So $p(N | S) = p(N \cap S)/p(S) \leq 0.03/0.15 \leq 0.2.$

Comment: In real life, situations like that in the last part of the exercise are quite common: the available data does not suffice to determine a precise probability, but may be enough to put non-trivial lower and/or upper bounds on it.

When calculating or proving with conditional probabilities it is easy to forget the fact that they are only partial functions and proceed as if they are full ones. In other words, it is easy to neglect the fact that $p(A | B)$ is not defined in the limiting case that $p(B) = 0.$ This can lead to serious errors. At the same time, it is quite distracting to have to check out these limiting cases while trying to solve the problem – you can easily lose the thread of what you are doing. How can these competing demands of rigour and insight be met?

The best procedure is to run with the fox and bay with the hounds. Make a *first draft* of your calculation or proof without worrying much about the limiting case, i.e. as if conditional probability was always defined. When finished, go back and *check each step* to see that it is really correct, if needed adding a side-argument (usually quite trivial) to cover the limiting case. Usually, everything should go smoothly. But sometimes there may be surprises!

EXERCISE 6.4.4 (WITH PARTIAL SOLUTION)

Let $p: S \to [0,1]$ be a probability function and suppose that $p(B) \neq 0.$ Show the following.

(a) $p(A_1 \cup A_2 | B) = p(A_1 | B) + p(A_2 | B)$ whenever the sets $A_1 \cap B$, $A_2 \cap B$ are disjoint.

(b) $p(-A | B) = 1 - p(A | B).$

(c) If $p(A) \geq p(B)$ then $p(A | B) \geq p(B | A).$

(d) $p(A | B)/p(B | A) = p(B)/p(A)$, assuming that also $P(A)/P(B) \neq 0.$

Solution to (a–c):

(a) Suppose $A_1 \cap B$, $A_2 \cap B$ are disjoint. Then:

$$
\begin{aligned}
p(A_1 \cup A_2 | B) &= p((A_1 \cup A_2) \cap B))/p(B) \text{ by definition of conditional} \\
&\qquad \text{probability} \\
&= p((A_1 \cap B) \cup (A_2 \cap B))/p(B) \text{ by Boolean distribution in} \\
&\qquad \text{numerator} \\
&= (p(A_1 \cap B) + p(A_2 \cap B))/p(B) \text{ using disjointedness} \\
&\qquad \text{assumption} \\
&= p(A_1 \cap B)/p(B) + p(A_2 \cap B)/p(B) \text{ by arithmetic} \\
&= p(A_1 | B) + p(A_2 | B) \text{ by definition of conditional} \\
&\qquad \text{probability}
\end{aligned}
$$

(b) $p(-A|B) = p(-A \cap B)/p(B)$ by definition of conditional probability
$$
\begin{aligned}
&= (p(-A \cap B) + p(A \cap B) - p(A \cap B))/p(B) \text{ by arithmetic} \\
&= (p(-A \cap B) \cup (A \cap B)) - p(A \cap B))/p(B) \text{ by disjointedness} \\
&= (p(B) - p(A \cap B))/p(B) \text{ by Boolean equality } B = (-A \cap B) \cup \\
&\quad (A \cap B) \\
&= 1 - [p(A \cap B))/p(B)] \text{ by arithmetic} \\
&= 1 - p(A|B) \text{ by definition of conditional probability.}
\end{aligned}
$$

(c) Suppose $p(A) \geq p(B)$. Then $p(A|B) = p(A \cap B)/p(B) \geq p(A \cap B)/p(A) = p(B|A)$, using the supposition $p(A) \geq p(B)$ and initial hypothesis $p(B) \neq 0$ to ensure that all terms are well-defined and to get the middle $\geq$.

Alice Box: What is $A|B$?

Alice: I understand the definitions and can do the exercises. But there is a question bothering me. *What is $A|B$?* The definition of conditional probability tells me how to calculate $p(A|B)$, but it does not say what $A|B$ itself is. Is it a subset of the sample space S like A,B, or some new kind of 'conditional object'?

Hatter: Neither one nor the other! Don't be misled by notation: $A|B$ is just a traditional shorthand way for writing the *ordered pair* (A,B). The vertical stroke is there merely to remind you that there is a division in the definition. Conditional probability is thus a *partial two-place function*, not a one-place function. A notation uniform with the rest of mathematics would be simply $p(A,B)$. The expression $A|B$ does not stand for a subset of the sample set S. It is an evocative but potentially misleading way of referring to the ordered pair (A,B).

(Continued)

Alice Box: (Continued)

Alice: Does an expression like $p(A|(B|C))$ or $p((A|B)|C)$ mean anything?

Hatter: Nothing whatsoever! Try to unpack the second one, say, according to the definition of conditional probability. We should have $p((A|B)|C) = p((A|B) \cap C)/p(C)$. The denominator is OK, but the numerator is meaningless, for $p(X \cap C)$ is defined only when X is a subset of the sample set S, and $A|B$ doesn't fit the bill.

Alice: So the operation of conditionalization can't be iterated?

Hatter: Not so fast! We *can* iterate conditionalization, but not like that! Back to the text to explain how.

As we have emphasized, a probability function p: $\mathcal{P}(S) \to [0,1]$ over a sample space S has only one argument, while its associated conditional probability function has two; it is a partial two-place function with domain $\mathcal{P}(S) \times \{X \subseteq S: p(X) \neq 0\}$.

But like all two-place functions, full or partial, a conditional probability function can be transformed into a family of one-place functions by the operation of projection (Section 3.5) on left or right argument, and in this case the interesting projections are on the right. For every probability function p: $S \to [0,1]$ and every $B \subseteq S$ with $p(B) \neq 0$ we have the one-place function p_B defined by putting $p_B(A) = p(A|B) = p(A \cap B)/p(B)$ for all $A \subseteq S$. The reason why these projections are of interest is that each such function p_B: $S \to [0,1]$ is *also* a probability function over the same sample space S, as is easily verified. It is called the *conditionalization* of p on B.

Since the conditionalization of a probability function is also a probability function, there is a perfectly good sense in which the operation of conditionalization may be iterated. Although Alice's expressions $p(A|(B|C))$ and $p((A|B)|C)$ are meaningless, the functions $(p_B)_C$ and $(p_C)_B$ are well-defined when $p(B \cap C) \neq 0$. Moreover, it is not difficult to show that the order of conditionalization is immaterial, and that every iterated conditionalization collapses into a suitable one-shot conditionalization. Specifically: $(p_B)_C = p_{B \cap C} = p_{C \cap B} = (p_C)_B$ when all these functions are well-defined, i.e. when $p(B \cap C) \neq 0$.

EXERCISE 6.4.5 (WITH PARTIAL SOLUTION)

Let p: $S \to [0,1]$ be a probability function. Show the following.

(a) $p = p_S$.

(b) If $p(B) \neq 0$, then $p_B(A_1 \cup A_2) = p_B(A_1) + p_B(A_2)$ whenever the sets $A_1 \cap B$, $A_2 \cap B$ are disjoint.

(c) If $p(B) \neq 0$, then $p_B(-A) = 1 - p_B(A)$.

(d) If $p(B \cap C) \neq 0$ then $(p_B)_C = p_{B \cap C} = p_{C \cap B} = (p_C)_B$ as claimed in the text.

Solutions and hints for (a–c):

(a) By definition, $p_S(A) = p(A \cap S)/p(S) = p(A)/1 = p(A)$.

(b), (c) Use Exercise 6.4.4.

6.5 Interlude: Simpson's Paradox

We pause for a moment to note a surprising fact known as *Simpson's paradox* (although it was in effect remarked as early as 1903 by Yule, and even earlier in 1899 by Pearson). It may be stated in terms of ratios or in terms of probabilities, and it shows – just in case you hadn't already noticed – that even elementary discrete probability theory is full of traps for the unwary.

Suppose we have a sample space S, a partition of S into two cells F, $-F$ and another partition of S into two cells C, $-C$. To fix ideas, let S be the set of all applicants for vacant jobs in a university in a given period. It is assumed that all applicants are female or male (but not both) and that every applicant applies to either computer science or mathematics (but not both). We write F, $-F$ for the female and non-female (i.e. male) applicants respectively; and C, $-C$ are the applicants to the departments of computer science and elsewhere (i.e. mathematics). The four sets $F \cap C$, $F \cap -C$, $-F \cap C$, $-F \cap -C$ form a partition of S (check it). Now let H be the set of applicants actually hired. Suppose that:

$$p(H|F \cap C) > P(H| -F \cap C)$$

i.e. the probability that an arbitrarily chosen female candidate to the computer science department is hired, is greater than the probability that an arbitrarily chosen male candidate to the computer science department is hired. Suppose that the same inequality holds regarding the mathematics department, i.e. that

$$p(H|F \cap -C) > P(H| -F \cap -C).$$

Intuitively, it seems inevitable that the same inequality will hold when we take unions on each side, i.e. that

$$p(H|F) > P(H| -F).$$

But in fact it need not do so! We give a counter-example (based on a law suit that was actually brought against a US university). Let S have 26 elements, neatly partitioned into 13 female and 13 male. Eight women apply to computer science, of which two are hired, while five men apply to the same department, of which one is hired. At the same time, five women apply to mathematics, of which four are hired, while eight men apply and six are hired. We calculate the probabilities.

$$p(H|F \cap C) = p(H \cap F \cap C)/p(F \cap C) = 2/8 = 0.25 \text{ while}$$
$$p(H|-F \cap C) = p(H \cap -F \cap C)/p(-F \cap C) = 1/5 = 0.2$$

so that the first inequality holds. Likewise

$$p(H|F \cap -C) = p(H \cap F \cap -C)/p(F \cap -C) = 4/5 = 0.8 \text{ while}$$
$$p(H|-F \cap -C) = p(H \cap -F \cap -C)/p(-F \cap -C) = 6/8 = 0.75$$

and thus the second inequality also holds. But we have:

$$p(H|F) = p(H \cap F)/p(F) = (2+4)/13 = 6/13 < 0.5$$
$$p(H|-F) = p(H \cap -F)/p(-F) = (1+6)/13 = 7/13 > 0.5$$

and so the third inequality fails – giving rise to the lawsuit!

Of course, the 'paradox' is not a paradox in the strict sense of the term: it is not a contradiction arising from apparently unquestionable principles. It does not shake the foundations of probability theory. But it shows that there can be surprises even in elementary discrete probability, as also in the theory of comparative ratios on which it rests. Intuition is not always a reliable guide in matters of probability, and even experts can go wrong.

6.6 Independence

There are two equivalent ways of defining the independence of two events. One uses conditional probability, the other is in terms of multiplication. We begin with the route via conditional probability.

Let $p: \mathcal{P}(S) \to [0,1]$ be any probability function on a (finite) sample space S. Let A,B be events (i.e. subsets of S). We say that A *is independent of* B (modulo p) iff $p(A) = p(A|B)$ (in the principal case) or (limiting case) $p(B) = 0$. In other words, iff conditionalizing on B makes no difference to the probability of A. In the language of projections, iff $p(A) = p_B(A)$ when p_B is defined.

Note again that this is just mathematics, and not a matter of the presence or absence of causality. Remember too that independence is modulo the probability function chosen: A may be independent of B with respect to one function p, but not so with respect to another probability function p'.

The other definition, in terms of multiplication, is as follows: A is independent of B iff $p(A \cap B) = p(A) \cdot p(B)$. The two definitions are equivalent. Some people find the definition in terms of conditional probabilities more intuitive, except for its limiting case, which is a little annoying. But the definition in terms of multiplication is usually easier to work with, and we adopt it as our official definition. In general, exercises should be solved using it.

EXERCISE 6.6.1 (WITH PARTIAL SOLUTION)

(a) Show the equivalence of the two definitions of independence.

(b) Show that the relation of independence is symmetric.

(c) Show that A is independent of B in each of the four limiting cases that $p(B) = 0$, $p(B) = 1$, $p(A) = 0$, $p(A) = 1$.

(d) Show that if A, B are disjoint, A is independent of B only if $p(A) = 0$ or $p(B) = 0$.

Solution to (a): Suppose first that $p(A \cap B) = p(A) \cdot p(B)$ and $p(B) \neq 0$. Then we may divide both sides by $p(B)$, giving us $p(A) = p(A \cap B)/p(B) = p(A|B)$ as desired.

Conversely, suppose either $p(B) = 0$ or $p(A) = p(A|B)$. In the former case $p(A \cap B) = 0 = p(A) \cdot p(B)$ as needed. In the latter case $p(A) = p(A|B) = p(A \cap B)/p(B)$ so multiplying both sides by $p(B)$ we have $p(A \cap B) = p(A) \cdot p(B)$ and we are done.

One advantage of the multiplicative definition is that it generalizes elegantly to the independence of any finite collection of events. Let $A_1, \ldots, A_n$ $(n \geq 1)$ be any such collection. We say that they are *independent* iff for every m with $1 \leq m \leq n$ we have: $p(A_1 \cap \ldots \cap A_m) = p(A_1) \cdot \ldots \cdot p(A_m)$. Equivalently, using a definition that may at first glance seem circular, but is in fact recursive (see Chapter 4): iff every proper non-empty subcollection of $\{A_1, \ldots A_n\}$ is independent and $p(A_1 \cap \ldots \cap A_n) = p(A_1) \cdot \ldots \cdot p(A_n)$. It should be emphasized that the independence of n events does not follow from their pairwise independence; a counterexample will emerge in one of the exercises.

When it holds, independence is a powerful tool for calculating probabilities. We illustrate this with an example. Five percent of computers sold by a store are defective. Today the store sold three computers. Assuming that these three form a random sample, with also independence as regards defects, what are (i) the probability that all of them are defective, (ii) the probability that none of them have defects, (iii) the probability that at least one is defective?

We are told that the probability $p(D_i)$ of defect for each computer c_i ($i \leq 3$) taken individually is 0.05, and that these probabilities are independent. So (i) the probability $p(D_1 \cap D_2 \cap D_3)$ that all three are defective is $(0.05)^3 = 0.000125$. (ii) The probability $p(-D_i)$ that a given computer is not defective is 0.95, so by independence the probability $p(-D_1 \cap -D_2 \cap -D_3)$ that none are defective is $(0.95)^3 = 0.857375$, considerable less than 0.95.

It remains to answer part (iii), the probability that at least one is defective. The quickest way of doing this is by noticing that at least one is defective iff it is not the case that none of them is defective. That is, $D_1 \cup D_2 \cup D_3 = -(-D_1 \cap -D_2 \cap -D_3)$ by de Morgan. We already have $p(-D_1 \cap -D_2 \cap -D_3) = (0.95)^3 = 0.857375$, so $p(D_1 \cup D_2 \cup D_3) = 1 - 0.857375 = 0.142625$.

Alice Box: Looking closely at the example

Alice: You said 'the quickest way'. Is there another one?

Hatter: Yes there is. Instead of de Morgan, we can use the fact that $p(D_1 \cup D_2 \cup D_3) = p(D_1) + p(D_2) + p(D_3) - p(D_1 \cap D_2) - p(D_1 \cap D_3) - p(D_2 \cap D_3) + p(D_1 \cap D_2 \cap D_3)$. In the problem we are given the first three values, and can obtain the remaining four by multiplication on the assumption of independence. This is rather tedious with three events (and much more so when for larger n), although it is quicker when there are only two, since $p(D_1 \cup D_2) = p(D_1) + p(D_2) - p(D_1 \cap D_2)$. So in general, don't use it when there are more than two events to join; use the de Morgan approach instead.

Alice: OK, but I think that there is a flaw in the de Morgan argument. In the statement of the problem, it was given that the sets D_i are independent. But if we look closely at the calculation of $p(-D_1 \cap -D_2 \cap -D_3)$, we see that it assumes that the complement sets $-D_i$ are independent.

Hatter: Flaw, no; but gap, yes! In fact, if the D_i are independent, so are the $-D_i$. For the case $n = 2$, that will be part of the next exercise.

EXERCISE 6.6.2 (WITH PARTIAL SOLUTION)

(a) Show that the relation of independence between two events is not reflexive, indeed, that it is 'almost irreflexive' in the sense that A is independent of A only in a very limiting case. Specifically, A is independent of A iff $p(A) \in \{1,0\}$.

(b) Use the above and symmetry to show that the relation of independence is not transitive.

(c) Show that the following four conditions are all equivalent (modulo any given probability function):

A and B are independent

A and $-B$ are independent

$-A$ and B are independent

$-A$ and $-B$ are independent

Solution to (c): We begin by showing the equivalence of the first to the second, and sketch the remainder. Suppose $p(A \cap B) = p(A) \cdot p(B)$; we want to show that $p(A \cap -B) = p(A) \cdot p(-B)$. Now $p(A) = p(A \cap B) + p(A \cap -B) = p(A) \cdot p(B) + p(A \cap -B)$ so $p(A \cap -B) = p(A) - p(A) \cdot p(B) = p(A) - p(A) \cdot (1 - p(-B)) = p(A) - (p(A) - p(A) \cdot p(-B)) = p(A) \cdot p(-B)$ as desired.

Given this, we can get the equivalence of the first to the third by symmetry and the established result. Then we get the equivalence of the second to the fourth by the same means.

The usefulness of independence in making calculations should not, however, blind us to the fact that in real life it holds only exceptionally, and should not be assumed without good reason. Examples abound. In one from recent UK legal history, an expert witness cited the accepted probability of an infant cot death in a family, and obtained the probability of two successive such deaths by simply squaring the figure. This is legitimate only if we can safely assume that the two probabilities are independent, which is highly questionable. Nevertheless, even when independence fails, it can sometimes be useful to consider it in a search for lower or upper bounds for a probability.

EXERCISE 6.6.3

Two archers shoot at a target. The probability that A_1 gets a bulls-eye is 0.2, the probability that A_2 gets a bulls-eye is 0.3. Assuming independence, what is (i) the probability that they both get a bulls-eye?

(ii) That neither gets a bulls-eye? (iii) Assuming also independence for successive tries, that neither gets a bulls-eye in ten successive attempts each?

6.7 Bayes' Theorem

We have emphasised that conditional probability does not commute: in general, $p(A|B) \neq p(B|A)$. But there are situations in which we can calculate the value of one from the other, provided we are given some supplementary information. In this section we see how this is done.

Assume that $p(A)$, $p(B)$ are non-zero. By definition, $p(A|B) = p(A \cap B)/p(B)$, so $p(A \cap B) = p(A|B) \cdot p(B)$. But likewise $p(B \cap A) = p(B|A) \cdot p(A)$. Since $A \cap B = B \cap A$ this gives us $p(A|B) \cdot p(B) = p(B|A) \cdot p(A)$ and so finally:

$$p(A|B) = p(B|A) \cdot p(A)/p(B).$$

Thus we can calculate the conditional probability in one direction if we know it in the other direction and also know the two unconditional (also known as *prior*, or *base rate*) probabilities.

We can go further. Since $p(B) = p(B|A) \cdot p(A) + p(B|-A) \cdot p(-A)$ – see the exercise below – we can substitute in the denominator to get the following equation, which is a special case of what is known as *Bayes' theorem*:

Assuming that $p(A), p(B)$ are non-zero, so $p(A|B)$ and $p(B|A)$ are well-defined, $p(A|B) = p(B|A) \cdot p(A)/p(B|A) \cdot p(A) + p(B| - A) \cdot p(-A)$.

So we can calculate the probability of $p(A|B)$ provided we know both of the conditional probabilities $p(B|A)$ and $p(B|-A)$ in the reverse direction, and the unconditional probability $p(A)$ (from which we can of course get $p(-A)$).

The equation is named after the eighteenth century clergyman-mathematician who first noted it, and it is surprisingly useful. For it can often happen that we have no information permitting us to determine $p(A|B)$ directly, but do have statistical information, i.e. records of frequencies, that permit us to give figures for $p(B|A)$ and for $p(B|-A)$, together with some kind of estimate of $p(A)$. With that information, Bayes' theorem tells us how to calculate $p(A|B)$.

EXERCISE 6.7.1

Verify the claim made in the text above, that $p(B) = p(B|A) \cdot p(A) + p(B|-A) \cdot p(-A)$. *Hint:* Use the fact that $B = (B \cap A) \cup (B \cap -A)$.

Warning: Applications of Bayes' theorem can be controversial. The figures coming out of the calculation are no better than those going into it – as the saying goes, garbage in, garbage out. In practice, it often happens that we have good statistics for estimating $p(B|A)$, perhaps less reliable ones for $p(B|-A)$, and only vague theoretical reasons for guessing approximately at $p(A)$. A good critical sense is needed to avoid being led into fantasy.

Indeed, probability theorists and statisticians tend to be divided into what are called *Bayesians* and their opponents. The difference is one of attitude or philosophy rather than mathematics. Bayesians tend to be happy with applying Bayes' theorem to get a value for $p(A|B)$ even in contexts where available statistics give us no serious indication of the value of the 'prior' $p(A)$. They are willing to rely on non-statistical reasons for its estimation or confident that possible errors deriving from this source can, by sophisticated techniques, be taken into account and even 'washed out' of the system. Their opponents tend to be much more cautious, unwilling to assume a probability that does not have a serious statistical backing. As one would expect, subjectivists tend to be Bayesian, while frequentists do not.

It is beyond the scope of this book to enter into the debate, although the reader can presumably guess where the author's sympathies lie. And it should be remembered that whatever position one takes in the debate on applications, Bayes' theorem as a mathematical result, stated above, remains a provable fact.

EXERCISE 6.7.2 (WITH SOLUTION)

At any one time, approximately 15% of patients in a ward suffer from the HIV virus. Further, of those that have the virus, about 90% react positively on a certain test, whereas only 3% of those lacking the virus react positively. (i) Find the probability that a patient has the virus given that the test result is positive. (ii) What would that probability be if 60% of the patients in the ward suffered from the HIV virus, the other data remaining unchanged?

Solution: First, we translate the data and goal into mathematical language. The sample set S is the set of all patients in the ward. Write V for the set of those with the HIV virus and P for the set of those who react positive. We are given the probabilities $p(V) = 0.15$, $p(P|V) = 0.9$, $p(P|-V) = 0.03$. For (i) we are asked to find the value of $p(V|P)$. Bayes' theorem tells us that $p(V|P) = p(P|V) \cdot p(V) / p(P|V) \cdot p(V) + p(P|-V) \cdot p(-V)$. The rest is calculation: $p(V|P) = 0.9 \cdot 0.15 / (0.9 \cdot 0.15 + 0.03 \cdot 0.85) = 0.841$ to the first three decimal places. For (ii) we are given $p(V) = 0.6$ with the other data unchanged, so this time $p(V|P) = 0.9 \cdot 0.6 / (0.9 \cdot 0.6 + 0.03 \cdot 0.4) = 0.978$ to the first three decimal places.

Comment: Notice how the solution depends on the value of the unconditional probability (alias base rate, or prior) $p(V)$. Other things being equal, a higher prior gives a higher posterior probability: as $p(V)$ moved from 0.15 to 0.6, $p(V|P)$ moved from 0.841 to 0.948.

Bayes' theorem can be generalized. Clearly, the pair $A, -A$ gives us a two-cell partition of the sample space S (complement being taken as relative to that space), and it is natural to consider more generally any partition into n cells $A_1, \ldots, A_n$. Essentially the same argument as in the two-cell case then gives us the following *general version of Bayes' theorem*. It should be committed to memory:

Consider any event B and partition $\{A_1, \ldots, A_n\}$ of the sample space, all with non-zero probability. Then for each $i \leq n$ we have:

$$p(A_i|B) = p(B|A_i) \cdot p(A_i) / \sum_{j \leq n}[p(B|A_j) \cdot p(A_j)].$$

EXERCISE 6.7.3

A telephone helpline has three operators, Alfred, Betty, Clarice. They receive 33, 37, 30 percent of the calls respectively. They manage to solve the problems of 60, 70, 90 percent of their respective calls. What is the probability that a successfully handled problem was dealt with by Alfred, Betty, Claire respectively? *Hint*: First translate into mathematical language and then use Bayes' theorem with $n = 3$.

6.8 Random Variables and Expected Values

In this section we explain the notion traditionally called a 'random variable', also known as a payoff function, and its expected value given a probability function. We then define the important structural concept of an induced probability function, and use it to see expected values from another angle.

6.8.1 Random Variables

Let $p: S \to [0,1]$ be any probability distribution on a (finite) sample space S. Suppose that we have another function f giving some kind of magnitude or value, positive or negative, for each of the elements of S. For example, S may consist of runners in a horse-race, and f may give the amount of money that that will be won or lost, under some array of bets, when a given horse wins. Or S could consist of the outcomes of throwing a die twice, with f specifying say the sum of the numbers appearing in the two throws. All we need to assume about f is that $f\colon S \to \mathbf{R}$, i.e. that it is on S and into the reals. In probability theory, such a function $f\colon S \to \mathbf{R}$ is traditionally called a *random variable*; in decision theory it is often called a *payoff function*.

Alice Box: Random? variable?

Alice: That's a very strange term, 'random variable'. In the language that we have been learning, a variable is something syntactic – a letter ranging over a domain. And in the first part of this chapter, when we spoke of a random element of a set we meant one serving as argument of an equiprobability distribution. This is neither! So why use the term?

Hatter: The language of probability, like that of counting, goes back beyond the era of modern set theory. It persists – some would say stubbornly – to this day. Table 6.1 at the beginning of this chapter gives a translation key from modern into traditional language. In particular:

- Functions $f\colon S \to \mathbf{R}$ are traditionally called *random variables*, and written X.

- The image $f(S)$ of a random variable is called its *range space*, written R_X.

- The composition function $p \circ f^{-1}$ is called the *distribution of* the random variable X.

- The probability $p \circ f^{-1}(x)$ i.e. $p(\{s \in S\colon f(s) = x\})$ is written $p(X = x)$.

Alice: Do I have to remember this?

Hatter: Well, you should at least keep it handy for reference. Life would be easier if everyone were to make use of the terminology and notation of the language of sets and functions. But most books on probability theory – even most introductory sketches intended for students of computer science – are still expressed in the traditional manner. So if you want to read such books, you have to learn the language.

6.8.2 Expectation

Is there any way in which we can bring a probability distribution $p\colon S \to [0,1]$ and a payoff function (alias random variable) $f\colon S \to \mathbf{R}$ to work together? The natural idea is to weigh one by the other. A low payoff with a high probability may be as desirable than a large payoff with small probability. Thus we may consider the *probability-weighted payoff function* (equivalently, payoff-weighted probability distribution) to be the function $f \cdot p \colon S \to \mathbf{R}$ defined by putting $(f \cdot p)(s) = f(s) \cdot p(s)$. Note that in the expression $f \cdot p$, the dot does not stand for composition; it is a two-place operation on functions with numbers as values, known as *point-wise multiplication*.

Example: Suppose that S consists of the horses in a race, $p(s)$ is the probability that horse s will win (as determined from, say, its track record) and $f(s)$ is the amount that you stand to gain or lose from your bets if s wins. Then $f(s) \cdot p(s)$ is that gain or loss weighted by its probability.

This in turn leads naturally to the concept of *expected value* or, more briefly, *expectation*. It is a probability-weighted average. We introduce it through an example of a three-horse race. Let $S = \{a, b, c\}$ be the runners, with probabilities $p(a) = 0.1$, $p(b) = 0.3$, $p(c) = 0.6$. Suppose the payoff function puts $f(a) = 12$, $f(b) = -1$, $f(c) = -1$. Then the expectation of f given p is the sum $f(a) \cdot p(a) + f(b) \cdot p(b) + f(c) \cdot p(c) = 12(0.1) - 1(0.3) - 1(0.6) = 0.3$.

In general terms, given a probability distribution $p \colon S \to [0,1]$ and a payoff function (alias random variable) $f \colon S \to \mathbf{R}$, we define the *expectation* of f given p, written $\mu(f, p)$ or just μ, by the equation $\mu = \mu(f, p) = \Sigma_{s \in S}\{f(s) \cdot p(s)\}$.

EXERCISE 6.8.1

(a) In the example given, compare the expectation of f given p with the arithmetic mean (alias average) of values of f, calculated (without reference to probability) by the usual formula $mean(f) = \Sigma_{s \in S}\{f(s)\}/ \#(S)$. Explain the basic reason for the difference.

(b) Show that in the special case that $p \colon S \to [0,1]$ is an equiprobable distribution expectation equals mean.

When the expectation of a gamble comes out negative, then it is not a good bet to make – one is likely to lose money. When it is positive, it is favourable.

Alice Box: A good bet?

Alice: So in our horse-race example, where $\mu = 0.3$, it is a good bet to make?

Hatter: Well, yes.

(*Continued*)

Alice Box: (Continued)

Alice: But if I bet only once, then there only two possibilities: either I win or I lose. The probability of winning is low (0.1), while the probability of losing is high (0.9). So, even though the payoff on winning is high ($12) and on losing is low ($−1), I am much more likely to lose than to win. Not very attractive!

Hatter: If you bet only once, yes. But if you make the same bet many times, then it becomes highly probable that the outcome is close to the expected value. We have not shown that here, but it can be proven. In that case the bet becomes attractive.

Alice: Provided I have enough capital to be able to put up with some probable initial losses, without undue hardship. . .

Hatter: Very true! What this illustrates is that the mathematics of probability theory is one thing, its application another. The former is relatively straight-forward, the latter can be tricky.

6.8.3 Induced Probability Distributions

So far, we have been looking at things from the point of view of the sample space S. But it is also useful to consider them from the point of view of its image $f(S)$ under the payoff function. This is particularly so when $f(S)$ is smaller than S, which can happen when f is not injective.

We will generalize a little, considering any function $f\colon S \to U$ on the sample space S into an arbitrary set U. Typically U will be the set $\mathbf{R}$ of reals, and thus f will be a payoff function (alias random variable), but the definition that we are about to give makes sense for any choice of U.

There is a natural way in which $f\colon S \to U$ and $p\colon \mathrm{P}(S) \to [0,1]$ work together to determine an *induced probability distribution* on the range $f(S) \subseteq U$ into [0,1]. For each $u \in f(S)$, we take the inverse image $f^{-1}(\{u\}) = \{s \in S\colon f(s) = u\}$ of U, and then apply the probability function to that. In other words, we consider the function $p\circ f^{-1}$ on $f(S)$ into the real interval [0,1] defined by the equation: $p\circ f^{-1}(u) = p(\{s \in S\colon f(s) = u\})$.

The function $p\circ f^{-1}\colon f(S) \to [0,1]$ is thus the composition of the inverse of f with the probability function p. We are abusing notation just a little here. As we saw in the chapter on functions, the inverse of a function $f\colon S \to U$ is a relation from $f(S)$ back to S, but will not be a function into S unless f is injective. However, if we rise one level of abstraction and regard the value of f^{-1}, for a given $u \in f(S) \subseteq U$, to be *the set of all s ∈ S such that* $f(s) = u$, then this is always a function on $f(S)$ into the power set $\mathcal{P}(S)$ of S, and we may as well write as it briefly with the same notation $f^{-1}: f(S) \to \mathcal{P}(S)$. The following diagram may help visualise what is going on.

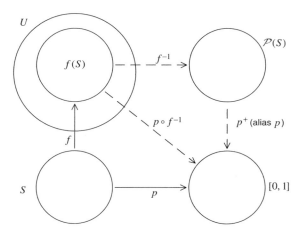

Figure 6.1   *Induced probability distribution.*

The interesting thing is that $p \circ f^{-1}: f(S) \to [0,1]$ turns out itself to be a probability distribution, but this time on $f(S) \subseteq U$ rather than on S. We call it the distribution on $f(S)$ *induced from p by f* or, for brevity, simply the *induced distribution*. In turn, it thus determines a probability function on $\mathrm{P}(f(S))$ into $[0,1]$, which we also call $p \circ f^{-1}$.

EXERCISE 6.8.2

    (a) If we identify the elements of S with their singletons, then Figure 6.1 may be simplified by treating S as a subset of $\mathcal{P}(S)$. Draw the diagram in this way, taking care to distinguish full from dotted arrows and to place correctly the beginning and end of each arrow. Keep a clean copy for reference.

    (b) Verify that, as claimed, $p \circ f^{-1}$ is a probability distribution on $f(S)$.

6.8.4 Expectation Expressed Using Induced Probability Functions

For any random variable $f\colon S \to \mathbf{R}$, we defined its expectation (given a probability function $p\colon S \to [0,1]$) by the equation $\mu = \mu(f,p) = \Sigma_{s \in S}\{f(s) \cdot p(s)\}$. This definition may be expressed equivalently in the language of induced probability functions, giving a rather deeper view of what is going on: $\mu = \mu(f,p) = \Sigma_{x \in f(S)}\{p \circ f^{-1}(x) \cdot x\}$.

Example: We review the horse-race example from the viewpoint of induced probability functions. Recall that $S = \{a,b,c\}$, with probabilities $p(a) = 0.1$, $p(b) = 0.3$, $p(c) = 0.6$ and payoffs $f(a) = 12$, $f(b) = -1$, $f(c) = -1$. In the table below:

- The top row gives the elements of the range space $f(S) = \{12, -1\}$. Note that although the sample space $S = \{a,b,c\}$ has three elements, $f(S)$ has only two elements.

- The second row gives the resulting values of $p \circ f^{-1}(x)$. We have $f^{-1}(12) = \{a\}$ so $p \circ f^{-1}(12) = p(a) = 0.1$, while $f^{-1}(-1) = \{b,c\}$ so $p \circ f^{-1}(-1) = p(b)+p(c)= 0.9$. Note that since $p \circ f^{-1}$ is a distribution on $f(S)$, the entries in the second row add to 1.

- The third row gives the resulting values of $p \circ f^{-1}(x) \cdot x$. Their sum is 0.3, which is the value of $\mu(f,p)$.

Table 6.2 Calculating expected value in an example.

$x \in f(S)$	12	-1
$p \circ f^{-1}(x)$	0.1	0.9
$p \circ f^{-1}(x) \cdot x$	1.2	-0.9

Another example: We walk through another example, from the same point of view. Consider a pair of fair dice about to be thrown. We want to determine the expected value of the sum of the two outcomes. First, we need to specify the sample space S: the 36 ordered pairs $(a,b) \in \{1,..,6\}^2$. Next, we specify the random variable (payoff function) $f\colon S \to \mathbf{R}$: we put $f(a,b) = a+b$. Hence its range $f(S)$ (alias R_X) is the set $\{2,\ldots,12\}$. In the following Table 6.3:

- The top row contains the elements of the range space $f(S) = \{2,\ldots,12\}$. Note that although the sample space S has 36 elements, $f(S)$ has only 11.

- The second row gives the resulting value of $p \circ f^{-1}(x)$. Taking for instance $x = 6$, $f^{-1}(x)$ has 5 elements $(1,5)$, $(2,4)$, $(3,3)$, $(4,2)$, $(5,1)$ out of the total of 36 in S. Since the distribution p is assumed to be equiprobable, each element of S gets the same probability $1/36$, so for $x = 6$, $p \circ f^{-1}(x) = 5/36$. Note that since $p \circ f^{-1}$ is a distribution on $f(S)$, the entries in the second row add up to 1.

- The bottom row gives the resulting values of $p \circ f^{-1}(x) \cdot x$.

- In accordance with the formula for expectation, $\mu(f, p) = \Sigma_{x \in f(S)}\{p \circ f^{-1}(x) \cdot x\}$ i.e. the sum of the figures in the bottom row, which is 7.

Table 6.3 Calculating expected value in a second example.

$x \in f(S)$	2	3	4	5	6	7	8	9	10	11	12
$p \circ f^{-1}(x)$	1/36	2/36	3/36	4/36	5/36	6/36	5/36	4/36	3/36	2/36	1/36
$p \circ f^{-1}(x) \cdot x$	2/36	6/36	12/36	20/36	30/36	42/36	40/36	36/36	30/36	22/36	12/36

In this example, the value of μ coincides with that of the ordinary mean of $f(S)$, calculated as $\Sigma\{x \in f(S)\}/11 = 7$, but this is exceptional. The two coincide because, on the one hand p is an equiprobable distribution on S and, on the other hand, the inverse f^{-1} of the random variable f is symmetric around argument 7. Modify either of these features, and the expectation will differ from the mean.

EXERCISE 6.8.3

Consider a fair coin to be thrown five times. Treating heads as 1 and tails as 0, we want to look at the sum of the five tosses. Specify the sample set S, random variable f, and range space $f(S)$, and draw up a table analogous to that of the dice example above to determine the expected value.

In this chapter we have only scratched the surface of discrete probability theory. For example, as well as expectation μ, which expresses the idea of weighted average, statisticians need measures of the dispersion of items around the expectation point – the degree to which they bunch up around it or spread far out on each side. The first steps in that direction are the concepts of *variance* and *standard distribution*. We will not go further in this brief chapter – there are pointers to reading below.

FURTHER EXERCISES

6.1. *Finite probability spaces*

(a) Consider a loaded die, such that the numbers 1–5 are equally likely to appear, but 6 is twice as likely as any of the others. To what probability distribution does this correspond?

(b) Consider a loaded die in which the odd numbers are equally likely, the even numbers are also equally likely, but are three times more likely than the odd ones. What is the probability of getting a prime number?

(c) Consider a sample space S with n elements. How many probability distributions are there on this space in which $p(s) \in \{0,1\}$ for all $s \in S$? For such a probability distribution, formulate a criterion for $p(A) = 1$, where $A \subseteq S$.

(d) Show by induction that when $A_1, \ldots, A_n$ are pairwise disjoint, then $p(A_1 \cup \ldots \cup A_n) = p(A_1) + \ldots + p(A_n)$.

6.2. *Unconditional probabilities*

(a) Consider the probability distribution in part (a) of the preceding exercise. Suppose that you roll the die three times. What is the probability that you get at least one 6?

(b) You have a loaded die with probability distribution like that in part (a) of the preceding exercise, and your friend has one with probability distribution like that in part (b). You both roll. What is the probability that you both get the same number?

(c) John has to take a multiple-choice examination. There are 10 questions, each to be answered 'yes' or 'no'. Marks are counted simply by the number of correct answers, and the pass mark is 5. Since he has not studied at all, John decides to answer the questions randomly. What is the probability that he passes?

(d) Mary is taking another multiple-choice examination, with a different marking system. There are again 10 questions, each to be answered 'yes' or 'no', and the pass mark is again 5. But this time marks are counted by taking the number of correct answers and subtracting the number of incorrect ones. If a question is left unanswered, it is treated as incorrect. If Mary answers at random, what is the probability that she passes?

6.3. *Conditional Probability*

(a) In a sports club, 70% of the members play football, 30% swim, and 20% do both. What are (i) the probability that a randomly chosen member plays football, given that he/she swims? (ii) the converse probability?

(b) Two archers Alice and Betty shoot an arrow at a target. From their past records, we know that their probability of hitting the target are 1/4 and 1/5 respectively. We assume that their performances are independent of each other. If we learn that exactly one of them hit the target, what is the probability that it was Alice?

(c) Show that, as claimed in the text, the conditionalization of any probability function is also a probability function whenever it is well-defined.

6.4. *Independence*

(a) A fair coin is tossed three times. Consider the events A, B, C that the first toss gives tails, the second toss gives tails, and we get exactly two tails in a row. Specify the sample space, and identify the three events by enumerating their elements. Show that A and B are independent, and likewise A and C are independent, but B and C are not independent.

(b) Use results established in the text to show that A is independent of B iff $p(A|B) = p(A|{-}B)$.

(c) Show that if A is independent of each of two disjoint sets B, C then it is independent of $B \cup C$.

(d) Construct an example of a sample space S and three events A, B, C that are pairwise independent but not jointly so.

6.5. *Bayes' theorem*

(a) Suppose that the probability of a person getting flu is 0.3, that the probability of a person having been vaccinated against flu is 0.4, and that the probability of a person getting flu given vaccination is 0.2. What is the probability of a person being vaccinated given that the person has flu?

(b) At any one time, approximately 3% of drivers have a blood alcohol level over the legal limit. About 98% of those over the limit react positively on a breath test, but 7% of those not over the limit also react positively. Find (i) the probability that a driver is over the limit given that the breath test is positive; (ii) the probability that a driver is not over the limit given that the breath test is negative; (iii) and (iv) the same results in a neighbouring country where, unfortunately, 20% of drivers are over the limit.

6.6. *Random variables and expected values*

(a) Consider a box of 10 items, of which 4 are defective. A sample of three items (order immaterial, without replacement, i.e. mode $\mathbf{O}{-}\mathbf{R}{-}$ in the notation of the chapter on counting) is drawn at random. What is the probability that it has exactly one defective item? What is the expected number of defective items?

(b) In a television contest, the guest is presented with four envelopes from which to choose at random. They contain $1, $10, $100, $1,000 respectively. What is the expected gain?

(c) A loaded coin has $p(H) = 1/4$, $p(T) = 3/4$. It is tossed three times. Specify the corresponding sample space S. Let f be the function (random variable) on this sample space that gives the number of heads that appear. Specify the range space $f(S)$. Construct a table showing the induced probability distribution on $f(S)$. Calculate the expectation $\mu = \mu(f,p)$.

(d) Show the equivalence of the definition of expected value with its characterization in terms of an induced probability function.

Selected Reading

Not many introductory texts of discrete mathematics say more than a few words on probability. The following cover more or less the same ground as in this chapter, but with mainly traditional terminology:

Ralph P. Grimaldi *Discrete and Combinatorial Mathematics*. Addison Wesley, 2004 (fifth edition), Chapter 3.4–3.8.

Seymour Lipschutz and Marc Lipson *Discrete Mathematics*. McGraw Hill Schaum's Outline Series, 1997 (second edition), Chapter 7.

Kenneth Rosen *Discrete Mathematics and its Applications*. McGraw Hill, 2007 (sixth edition), Chapter 6.

For those wishing to go further, the authors of the second text have also written:

Seymour Lipschutz and Marc Lipson *Probability*. McGraw Hill Schaum's Outline Series, 2000 (second edition).

7
Squirrel Math: Trees

Chapter Outline

This chapter introduces a kind of structure that turns up everywhere in computer science – trees. We will be learning to speak their language – how to talk about their components, varieties and uses – more than proving things about them. The flavour of the chapter is thus rather different from that of the preceding one on probability: much more use of spatial intuition, rather less in the way of demonstration.

We begin by looking at trees in their most intuitive form – *rooted* (alias *directed*) trees – first of all quite naked, and then clothed with *labels* and finally *ordered*. Particular attention will be given to the case of binary trees and their use in search procedures. We then turn to *unrooted* (or *undirected*) trees and their application to *span* graphs. As always, we remain in the finite case.

7.1 My First Tree

Before defining the general concept, we give an example. Consider the structure presented in Figure 7.1 below.

This structure is a rooted tree. It consists of a set T of fifteen elements $a,...,o$, together with a relation over them. The relation is indicated by the arrows. The elements of the set are called *nodes*, or vertices; the arrows representing pairs in the relation are called *links* (alias arcs or edges). We note some features of the structure.

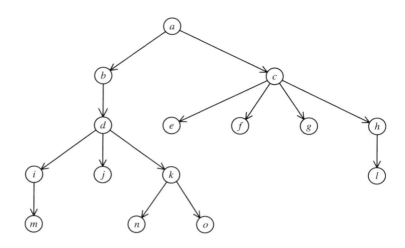

Figure 7.1   *Example of a rooted tree.*

- The relation is acyclic, and so in particular asymmetric and irreflexive (recall these notions from the chapter on relations).

- There is a distinguished element a of the set, from which all others may be reached by following paths in the direction of the arrows. In the language of sets, all other elements are in the transitive closure of $\{a\}$ under the relation. This is called the *root* of the tree, and it is unique.

- Paths may fork, continue, or stop. Thus from the node a we may pass to two nodes b and c. These are called the *children* of a. While a has just two children, d has three (i,j,k), and c has four (e,f,g,h). But b has only one (d), and m for example has no children. Mixing biological with botanical metaphors, nodes without children are called *leaves*. Any path from the root to a leaf is called a *branch* of the tree. Note that a branch is an entire path from root to a leaf – thus neither (b,d,i) nor (a,c,h) nor (a,b,k,n) is a branch of the tree in the figure.

- Paths never meet once they have diverged: we never have two or more arrows going to a node. In other words, each node has at most one *parent*. An immediate consequence of this and irreflexivity is that the link relation is intransitive.

Given these features, it is clear that in our example in Figure 7.1 we can leave the arrow-heads off the links, with the direction of the relation understood implicitly from the layout.

EXERCISE 7.1.1 (WITH PARTIAL SOLUTION)

This exercise concerns the tree of Figure 7.1. It is intended to sharpen intuitions before we give a formal definition of what a tree is.

(a) Identify all the leaves of the tree, and count them.

(b) Why are none of (b,d,i), (a,c,h), (a,b,k,n) branches of the tree?

(c) Identify all the branches of the tree and count them. Compare the result with that for leaves, and comment.

(d) The *link-length* of a branch is understood to be the number of links making it up. How does this relate to the number of nodes in the branch? Find the link-length of each branch in the tree.

(e) Suppose you delete the node m and the link leading into it. Would the resulting structure be a tree? What if you delete the node but leave the link? And if you delete the link but not the node?

(f) Suppose you add a link from m to n. Would the resulting structure be a tree? And if you add one from c to l?

(g) Suppose you add a node p without any links. Would you have a tree? If you add p with a link from b to p? And if you add p but with a link from p to b?

(h) Given the notions of parent and child in a tree, define those of sibling, descendant and ancestor in the natural way. Identify the siblings, descendants and ancestors of the node d in the tree.

Solutions to (a), (b), (e), (g):

(a) There are 8 leaves: m,j,n,o,e,f,g,l.

(b) Because the first does not reach a leaf, the second does not begin from the root, the third omits node d from the path that goes from the root to leaf n.

(e) If you delete node m and the arrow to it, the resulting structure is still a tree. But if you delete either alone, it will not be a tree.

(g) In the first case no, because the new node p is not reachable by a path from the root a; nor can it be considered a new root for the enlarged structure, because no other node is reachable from it. But in the second case, we would have a tree. In the third case we would not have a tree as p would not be reachable from a; nor would p be a new root for the enlarged structure, since a is not reachable from it.

7.2 Rooted Trees

There are two ways of defining the concept of a rooted tree. One is *explicit*, giving a necessary and sufficient condition for a structure to be a tree. The other is *recursive*, defining first the smallest rooted tree and then larger ones out of smaller ones. As usual, the recursive approach tends to be better for the computer, though not always for the human. Each way gives its special insight, and problems are sometimes more easily solved using one than another. We begin with the explicit approach.

For that, we need the notion of a *directed graph* or briefly *digraph*. This is simply any pair (G,R) where G is a set and R is a two-place relation over G. A rooted tree is a special kind of directed graph, but we can zoom in on it immediately, with no special knowledge of general graph theory. The only notion that we need that is not already available from Chapter 2 is that of a *path*. We used the notion informally in the preceding section, but we need a precise definition. If (G,R) is a directed graph, then a path is defined to be any finite sequence $a_0,\ldots,a_n$ ($n \geq 1$) of elements of G (not necessarily distinct from each other) such that each pair $(a_i,a_{i+1}) \in R$.

Note that in the definition of a path we require that $n \geq 1$; thus *we do not count empty or singleton sequences as paths.* Of course, they could also be counted if we wished, but that would tend to complicate formulations down the line, where we usually want to exclude the empty and singleton ones.

Note also that we do not require that all the a_i are distinct; thus (b,b) is a path when $(b,b) \in R$, and (a,b,a,b) is a path when both of $(a,b),(b,a) \in R$. Nevertheless, it will follow from the definition of a tree that paths like these (loops and cycles) never occur in trees.

Explicit definition of trees. A (finite) *rooted tree* is defined to be any finite directed graph (G,R) with an $a \in G$ (called the *root of the tree*) such that (i) for every $x \in G$ with $a \neq x$ there is a unique path from a to x but (ii) there is no path from a to a.

In addition, the *empty tree* (with both carrier set G and relation R empty) is often regarded as a rooted tree – despite the fact that it does not contain a root! This is evidently a limiting case, and is of little importance in the general theory, so much so that it is often excluded. But when we come to the special case of binary trees, it can facilitate recursive constructions.

When a graph (G,R) is a tree, we usually write its carrier set G as T, so that it is called (T,R). In this section, for brevity we will sometimes say simply *tree* for *rooted tree*; although when we get to unrooted trees towards the end of the chapter we will have to be more explicit. It is easy to establish some further properties of trees. Let (T,R) be any tree with root a (and so with T non-empty). Then:

- R is acyclic,

- a is the unique root of (T,R),

- a has no parent and every other element of G has just one parent,

- No two diverging paths ever meet.

We give informal proofs – all using proof by contradiction – and comment as we go. The proofs of acyclicity and parent numbers bring out the importance of the word 'unique' in the definition of a rooted tree.

- For *acyclicity*, recall from Chapter 2 that a relation R is said to be *acyclic* iff there is no path from any element x to itself, i.e. no path $x_0,\ldots,x_n$ ($n \geq 1$) with $x_0 = x_n$. Suppose that a tree fails acyclicity; we get a contradiction. By the supposition, there is a path from some element x to itself. But by the definition of a tree, there is also a unique path from the root a to x. Form the composite path made up of the path from the root to x followed by the one from x to itself. This is clearly another path from the root to x, and it must be distinct from the first path because it is longer. This contradicts uniqueness of the path from a to x.

- For *uniqueness of the root*, suppose for *reductio ad absurdum* that a and a' are distinct elements of T such that for every $x \in G$, if $a \neq x$ (resp. $a' \neq x$) there is a path from a to x (resp. from a' to x), but there is no path from a to a (resp. from a' to a'). From the first supposition there is a path from a to a', and by the second there is a path from a' to a. Putting these two paths together gives us one from a to a, giving us a contradiction.

- For the *parent numbers*, suppose that a has a parent x, so that $(x,a) \in R$. Since a is the root, there is a path from a to x. The composite path made up of this followed by the link (x,a) thus gives us a path from a to a: contradiction. Now let b be any element of T with $b \neq a$. By the definition of a tree there is a path from the root a to b, and clearly its last link gives us a parent of b. Now suppose that b has two distinct parents x and x'. Then there are paths from the root a to each of x and x', and compounding them with the additional links from x and x' to b gives us two distinct paths from a to b, contradicting the definition of a tree.

- Finally, if two *diverging paths* ever meet, then the node where they meet would have two distinct parents, contradicting the preceding property.

The *link-height* (alias *level*) of a node in a tree is defined recursively: that of the root is 0, and that of each of the children of a node is one greater than that of the node. The *node-height* of a node (alias just its *height* in many texts) is defined by the same recursion, except that the node-height of the root is set at 1. Thus for every node x, $node\text{-}height(x) = link\text{-}height(x) + 1$. As trees are usually drawn upside-down, the term 'depth' is often used instead of 'height'. Depending on the problem under consideration, either one of these two measures may be more convenient to use than the other. The *height* (whether in terms of nodes or links) *of the tree* itself may be defined to be the highest height (node/link respectively) of any of its nodes.

These notions are closely related to that of the length of a path in a tree. Formally, the *link-length* of a path $x_0,\ldots,x_n$ ($n \geq 1$) is n, its node-length is $n+1$. We can say that the height of a node (in terms of nodes or links) equals to the length (node/link respectively) of the unique path from the root to that node – except for the case of the root node, where the latter does not exist).

EXERCISE 7.2.1 (WITH PARTIAL SOLUTION)

(a) What is the (link/node)-height of the tree Exercise 7.1.1? Consider a set with $n \geq 2$ elements. What is the greatest possible (link/node)-height of any tree over that set? And the least possible?

(b) For $n = 4$, draw rooted trees with n nodes, one tree for each possible height.

(c) Show that the relation R of a rooted tree (T,R) is irreflexive and asymmetric, as those terms are defined in the chapter on relations.

(d) In Section 4.5 we gave a table of calls that need to be made when calculating $F(8)$ where F is the Fibonacci function. Rewrite this table as a tree, carrying its construction through to the elimination of all occurrences of F.

Solutions to (a) and (c):

(a) The link-height of the tree in Exercise 7.1.1 is 4, its node-height is 5. Let $n \geq 2$. The greatest possible (link/node)-height is obtained when the tree consists of a single branch, i.e. is a sequence of nodes. The node-length of such a branch is n, and its link-length is $n–1$. The least possible measure is obtained when the tree consists of the root with $n–1$ children; the link-length is then 1 and its node-length is 2.

(c) Irreflexivity and asymmetry both follow immediately from acyclicity (already established in the text) taking the cases $n = 1$ and $n = 2$.

First recursive definition. We pass now to the recursive definition of a
rooted tree. There are two main ways of doing this: the recursion step can bring
in a new root, or a new leaf. The most common presentation makes a new root,
and we give it first. In this context, it is convenient to make the root explicit in the
notation; thus a rooted tree is thought of as a triple (T, a, R) where a is the root.

- The basis of the definition stipulates that any singleton is a tree, with its
  element as root and with the empty relation for links. In other words, for
  any a, the triple $(\{a\}, a, \varnothing)$ is a rooted tree. In addition, if desired, the
  empty structure may be regarded as a rooted tree (but without a root).

- The recursion step tells us that whenever we have a finite collection of
  disjoint trees and a fresh item a, then we can form a new tree by taking a to
  be its root, linked to the roots of the given trees. Formally: Suppose
  (T_i, a_i, R_i) $(i \leq n)$ are trees with roots a_i. Suppose that the T_i are pairwise
  disjoint and $a \notin \cup\{T_i\}_{i \leq n}$. Then the structure $(\cup\{T_i \cup \{a\}\}_{i \leq n},\ a,$
  $\cup\{R_i \cup \{(a, a_i)\}_{i \leq n}\}$ is a tree with root a.

Note the conditions that the T_i must be disjoint and that a must be
fresh. The recursion step may be illustrated by the following diagram.

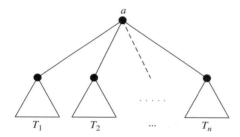

Figure 7.2   *Recursive definition of a tree.*

Here the T_i are represented schematically by triangles. They are the
immediate (proper) subtrees of the entire tree. They in turn may contain
more than a single node, and so have further subtrees, which will also be
subtrees of the whole.

In general, when (T, a, R) is any rooted tree with root a, then its *subtrees*
are the structures (T_b, b, R_b) where $b \in T$, T_b is the closure of $\{b\}$ under the
link relation R, and R_b is the restriction of R to T_b. Evidently, b is the root of
the subtree. As a limiting case, the empty tree may also be counted as a
subtree.

EXERCISE 7.2.2 (WITH PARTIAL SOLUTION)

(a) Use the above recursive definition of a tree to construct five trees of respective link-heights 0,1,2,3,4, each one obtained from its predecessor by an application of the recursion step.

(b) How many rooted subtrees does a non-empty rooted tree with $n \geq 1$ nodes have?

Solution to (b): Clearly there are a one-one correspondence between the nodes and the non-empty subtrees with them as roots. This includes the given tree, which is a subtree of itself. Thus there are n non-empty rooted subtrees. Hence there $n+1$ rooted subtrees including the empty one; n proper rooted subtrees including the empty one; and $n-1$ non-empty rooted proper subtrees.

We could also define rooted trees recursively by making fresh leaves. The basis, for the singleton tree, remains the same. The recursion step stipulates: Whenever we have a tree and a fresh item x, then we can form a new tree by linking any given node b of the old tree to x; the root of the new tree is the same as that of the old one. More formally: Suppose (T,a,R) is a rooted tree with root a, and let $b \in T$ and $x \notin T$. Then the structure $(T \cup \{x\}, a, \cup \{R \cup (b,x)\}$ is a rooted tree with the same root a.

Note that the new element x will be a leaf of the new tree; the node b may or may not have been a leaf of the old tree, but will not be a leaf of the new one. Non-leaves are often called *interior nodes*.

This recursive definition corresponds most closely to your thoughts and pencil strokes when you draw a tree on the page: you will usually start by placing a root, which remains so throughout the construction, adding new nodes by lengthening a branch (thus converting a leaf into an interior node) or dividing a branch. But the other recursive definition by 'new root' is the one most popular in computer science texts, as it lends itself most easily to large-scale computation.

EXERCISE 7.2.3

(a) Take the last tree that you constructed in the preceding exercise, and reconstruct it step by step using the 'new leaf' recursion.

(b) Use the 'new leaf' recursive definition to show that any tree with $n \geq 1$ nodes has $n-1$ links.

(c) Show the same directly from our explicit definition of a rooted tree and the properties that we established for it.

The three definitions of a rooted tree – explicit, recursive by 'new root' and recursive by 'new leaf' – are equivalent, but we will not prove that here, leaving it as a challenging problem at the end of the chapter.

What are some typical examples of trees in everyday life? The first that springs to mind may your family tree, but be careful! On the one hand, people have two parents, not one; and diverging branches of descendants can meet (legally as well as biologically) when cousins or more distant relations have children together. But if we make a family tree consisting of the male line only (or the female line only) of descendants of a given patriarch (resp. matriarch), then we do get a tree in the mathematical sense.

The most familiar example of a tree in today's world is given by the structure of folders and files that are available (or can be created) on your computer. Usually, these are constrained so as to form a tree. For those working as employees in an office, another familiar example of a tree is the staff organigram, in which the nodes are staff members (or rather, their functions) and the links indicate the immediate subordinates of each node in the hierarchy. In a well-constructed pattern of responsibilities, this will usually be a tree.

Trees also arise naturally whenever we investigate grammatical structure. In natural languages such as English, this reveals itself when we parse a sentence; in formal languages of computer science, mathematics and logic, parsing trees arise when we consider the syntactic structure of a formal expression.

Trees also arise in logic in other ways: a proof can be represented as a tree, with the conclusion as root and the assumptions as leaves. And one popular way for checking whether a formula of propositional logic is a tautology proceeds by constructing what is called its semantic decomposition tree.

We will give some examples of parsing trees shortly. Proof trees and semantic decomposition trees will be described in later chapters on logic. But first we should introduce some refinements into our theory, with the notions of *labelled* and *ordered* trees.

Alice Box: Which way up?

Alice: Before going on, why do you draw trees upside down, with the root at the top, as in Figure 7.1?

Hatter: Just convention.

Alice: Is that all?

Hatter: Well, that's the short answer, but not the whole story. The orientation also depends on how we are constructing the tree. If we are thinking of it as constructed

(*Continued*)

Alice Box: (Continued)

from the root to the leaves by a 'new leaf' recursion, then it is natural to write the root at the top of the page and construct downwards. As that is what humans most commonly do in small examples, diagrams are usually drawn with the root on top.

Alice: So when would we draw it the other way round?

Hatter: If we are considering the tree as built by a 'new root' recursion, so that we are proceeding from leaves to root, then it makes sense to put the leaves at the top of the page and work down to the root. For example, in logic texts, formal proofs are often represented by trees with the leaves (assumptions) at the top and the root (conclusion) below, while semantic decomposition trees are upside-down.

Alice: So it's a matter of convenience?

Hatter: Yes. The upside-down orientation is convenient when you are constructing the tree from root to leaves, but the right-way-up is natural when building from leaves to root.

7.3 Labelled Trees

In practice, we rarely work with naked trees. They are almost always clothed or decorated in some way – in the technical jargon *labelled*. Indeed, in many applications, the identity of the nodes themselves is of little or no importance; they are thought of as just 'points'; what is important are the labels attached to them.

A *labelled tree* is a tree (T,R) accompanied by a function $\lambda\colon T \to L$ (λ and L for 'label'). L can be any set, and the function λ can also be partial, i.e. defined on a subset of T. Given a node x, $\lambda(x)$ indicates some object or property that is placed at that point in the tree. Thus, heuristically, the nodes are thought of as hooks on which to hang labels. We know that the hooks are there, and that they are all distinct from each other, but what really interests us are the labels. In a diagram, we write the labels next to the nodes, and may even omit to give names to the nodes, leaving them as dots on the page.

It is important to remember that the labelling function *need not be injective*, and typically it will not be. For example, in a tree of folders and files in your computer, we may put labels indicating the number of subfolders immediately under them, or date of creation or of last modification. Different folders may thus have the same labels.

A given tree may come with several parallel systems for labelling its nodes. These could, of course, be amalgamated into one complex labelling function whose labels are ordered n-tuples of the component labels; but this is not always of any advantage. It is also sometimes useful to label the *links* rather than (or in addition to) the nodes. For example, we might want to represent some kind of distance, or cost of passage, between a parent and its children. Sometimes the location of the labels, on nodes or on links, is just a matter of convention and convenience. Remember that trees are tools for use, more than objects for study, and we have considerable freedom in adapting them to the needs of the task in hand.

An *ordered tree* is a tree in which the children of any node are put into a linear order and labelled with numbers $1,...,n$ to record the ordering. In this case the labelling function is a partial function on the tree (the root is not in its domain) into the positive integers. Alternatively, the labels could be put on the links, in which case it is a total function. In the important case of a binary tree (which we will consider in detail shortly) each node has at most two children; these are put in order and labelled 1,2 or, more commonly, 'left' and 'right'. We give two examples of labelled, ordered trees.

Example 1. Consider the arithmetic expression $(5+(2 \times x))-((y-3)+(2+x))$. Evidently, it is formed from two sub-expressions, $5+(2 \times x)$ and $(y-3)+(2+x)$ by applying the operation of subtraction. As this operation is not commutative, the order of application is important. We may represent the syntactic structure of the expression by a labelled tree, whose root is the whole expression, which has just two children, labelled by the two immediate sub-expressions. Continuing this process until decomposition is no longer possible, we get the following labelled, ordered tree.

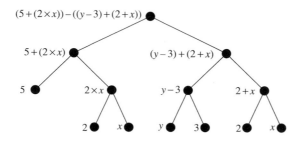

Figure 7.3   *Tree for syntactic structure of an arithmetic expression.*

Alice Box: Drawing trees

Alice: Why do we have to regard the expressions written alongside the nodes as *labels*? Why not just take them to be *the names* of the nodes?

Hatter: Because the labelling function need not be injective. In our example, we have two different nodes labelled by the numeral 2. We also have two different nodes

(*Continued*)

Alice Box: (Continued)

labelled by the variable x. They have to be distinguished because they have different roles in the expression and so different places in the tree

Alice: I see. But that leads me to another question. Where, in the figure, *are* the names of the nodes?

Hatter: We have not given them names. In the diagram they are simply represented by dots. To avoid clutter, we name them only if we really need to.

Alice: OK. One more question. As you said, in this expression order is important. That means that the children of each node must be given ordering labels, say 'left' and 'right'. Where are they?

Hatter: We let the sheet of paper take care of that. In other words, when we draw a diagram for an ordered tree, we follow the convention of writing the children in the required order from left to right on the paper. This is another way of reducing clutter.

Alice: So this tree is different from its mirror image, in which left and right are reversed on the page?

Hatter: Yes, since it is an ordered tree. If it were an unordered tree, they would be the same.

Actually, the tree in Figure 7.3 can be written more economically. The interior nodes need not be labelled with entire sub-expressions; we can just write their principal operations. This gives us the diagram in Figure 7.4. It does not lose information, for we know that the operation acts on the two children in the order given.

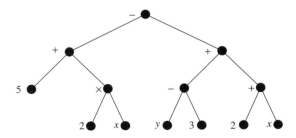

Figure 7.4 *Syntactic structure with economical labels.*

Syntactic decomposition trees such as that of Figure 7.3 are important for *evaluating* an expression, whether by hand or by computer. Suppose the variables

x and y are instantiated to 4 and 7 respectively. The most basic way of evaluating the expression as a whole is by 'climbing' the tree from leaves to root, always evaluating the children of a node before evaluating that node.

The tree in Figure 7.3 has a number of special structural properties, which are of considerable importance.

- Each node has at most two children. Such trees are called *2-trees*. In general, when each node of a tree has at most n children, it is called an *n*-tree.

- The tree is ordered in the sense defined above: the children of any given node are understood as having an order, expressed with labels or by a reading convention. An ordered 2-tree is called a *binary tree.* We will look at binary trees more closely in the next section.

- In fact, each node of the tree has either two children (in the case of the interior nodes) or no children (the leaves), so that no node has just one child. A binary tree with this property is sometimes called a *complete binary tree.*

- Note, however, that although the tree is complete in the sense defined, not all branches are of the same length.

Warning: Here as in so many parts of finite mathematics, terminology is protean. When reading any text, check exactly what it means by any of the above terms.

EXERCISE 7.3.1 (WITH PARTIAL ANSWER)

(a) Evaluate the expression labelling the root of Figure 7.3, for the values 4,7 for x,y respectively, by writing the values in the diagram alongside the labels.

(b) Is the sequence of steps that you made in the valuation uniquely determined, or could you also do it with a different sequence?

(c) Draw an example of a binary tree that is not a complete binary tree.

Answer to (b):

(b) Not quite uniquely determined. Although we must evaluate children before their parent, we can evaluate the children themselves in arbitrary order.

Example 2. Syntactic decomposition trees arise not only for arithmetic expressions, but for any kind of 'algebraic' expression in mathematics or logic. Take for example the formula of propositional logic $(p\rightarrow\neg q)\rightarrow(\neg r\leftrightarrow(s\wedge\neg p))$. Clearly, the structure of this expression can be represented by a syntactic decomposition tree.

EXERCISE 7.3.2 (WITH PARTIAL SOLUTION)

(a) Draw the syntactic decomposition tree for the formula of Example 2, once with full labelling and again with abbreviated labelling.

(b) Is it a 2-tree? A 2-tree? Is it a complete 2-tree? What are its node and link heights?

(c) Evaluate the formula for the values $p := 1$, $q := 1$, $r := 0$, $s := 0$, writing the values as additional labels next to the nodes on the tree. *Reminder:* You will need to use the truth tables given in logic boxes in Chapter 1.

Solution to (b): This is certainly a binary tree, since each node has at most two children. But to be annoyingly strict, whether it is a binary tree depends on how you drew it. For the two-place connectives $\rightarrow$, $\leftrightarrow$, $\wedge$ you branched left and right, and this can be read as an ordering convention. But what did you do with $\neg$? If the single link went straight down without a label, then no order is given. But if it went left, it can be read with that as label, and likewise for right. A rather pedantic distinction, but sometimes needed! Finally, even if presented as a binary tree, it is not a complete binary tree as there are nodes (in fact three of them) with exactly one child. Its node-height is 5, link-height 4.

7.4 Interlude: Parenthesis-Free Notation

In our arithmetical and logical examples of syntactic decomposition trees, we used brackets. This is necessary to avoid ambiguity. Thus the expression $5+2\times x$ is ambiguous unless brackets are put in (or a convention adopted, as indeed is usually done for multiplication, that it 'binds more strongly than', or 'has priority over' addition). Likewise, the logical expression $\neg r \leftrightarrow s \wedge \neg p$ is ambiguous unless parentheses are inserted or analogous conventions employed.

In practice, we frequently adopt priority conventions, and also drop brackets whenever different syntactic structures have the same semantic content, as for example with $x+y+z$. Such conventions help prevent visual overload and are part of the endless battle against the tyranny of notation.

For a while in the early twentieth century, a system of dots was used by some logicians (notably Whitehead and Russell in their celebrated *Principia Mathematica*) to replace brackets. Thus $(p \rightarrow \neg q) \rightarrow (\neg r \leftrightarrow (s \wedge \neg p))$ was written as $p \rightarrow \neg q. \rightarrow : \neg r \leftrightarrow .s \wedge \neg p$ where the dot after the q indicates a right bracket (the left one being omitted because it is at the beginning of the formula), the dot before the s marks a left bracket (its right partner omitted because it is at the end), and the two-dot colon marks another left bracket. The number of dots indicates the 'level' of the bracket. However the system did not catch on, and has now died out.

But is there any systematic way of doing without brackets altogether, without special priority conventions or alternative devices like dots? There is, and it was first noted by the philosopher/logician Łukasiewicz, early in the twentieth century. Instead of writing the operation *between* its arguments, just write it *before* them! Then brackets become redundant. For example, the arithmetical expression $(5+(2\times x))-((y-3)+(2+x))$ is written as $-+5\times 2x+-y3+2x$.

To decipher this last sequence, construct its (unique!) decomposition tree from the leaves to the root. The leaves will be labelled by the constants and variables. $+2x$ will label the parent of nodes labelled by 2 and x; $-y3$ will label the parent of nodes labelled by 3 and y, etc.

This is known as *Polish notation*, after the nationality of its inventor. There is also *reverse Polish* notation, where the operation is written after its arguments, and which likewise makes bracketing redundant. The ordinary way of writing expressions is usually called *infix* notation, as the operation is written in between the arguments. The terms *prefix* and *postfix* are often used for Polish and reverse Polish notation respectively.

EXERCISE 7.4.1 (WITH PARTIAL SOLUTION)

(a) Draw all possible decomposition trees for each of the ambiguous expressions $5+2\times x$ and $\neg r\leftrightarrow s\wedge\neg p$, writing the associated bracketed expression next to each tree.

(b) Draw the whole syntactic decomposition tree for the Polish expression $-+5\times 2x+-y3+2x$. Compare it with the tree for $(5+(2\times x))-((y-3)+(2+x))$.

(c) Put the propositional formula $(p\rightarrow\neg q)\rightarrow(\neg r\leftrightarrow(s\wedge\neg p))$ in both Polish (prefix) and reverse Polish (postfix) notation.

(d) Insert a few redundant brackets into the outputs just obtained, to make them friendlier to non-Polish readers.

Solution to (c) and (d):

(c) In prefix notation it is $\rightarrow\rightarrow p\neg q\leftrightarrow\neg r\wedge s\neg p$. In postfix notation it is $pq\neg\rightarrow r\neg sp\neg\wedge\leftrightarrow\rightarrow$.

(d) $\rightarrow(\rightarrow p\neg q)(\leftrightarrow(\neg r(\wedge s\neg p))$ and $(pq\neg\rightarrow)(r\neg(sp\neg\wedge)\leftrightarrow)\rightarrow$.

7.5 Binary Search Trees

Computer science and logic both love binary concepts. In the case of logic, they appear at its theoretical foundations: classical logic is two-valued, with truth-tables for the logical connectives defined using functions into the two-element set $\{0,1\}$. In computer science, they derive from a basic practical fact: a switch can be on or off. This means that an essential role is played by bits, and binary notions are propagated through the whole superstructure. Of all kinds of tree, the most intensively employed in computing is the binary one. So we look at it in more detail.

As introduced in the preceding section, a *binary tree* is a rooted tree (possibly empty) in which each node has at most two children, equipped with an ordering that labels each child in the tree (even only children) with a tag *left* or *right*. When drawing the tree on paper we need not write these labels explicitly; we can understanding them as given by the position of nodes on the page.

Example. Of the four trees diagrammed below, the first is not a binary tree, since the child of the root has not been given (explicitly or by orientation on the page) a left or right label. The remaining three are binary trees. The second and third are different binary trees, since the second gives a right label to the child of the root, while the third gives it a left label. The third and the fourth are the same when considered simply as binary trees, but they differ in that the fourth has additional (numerical) labels.

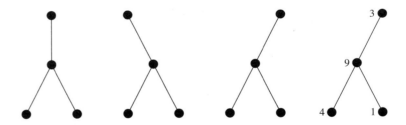

Figure 7.5   *Which are binary trees?*

What are binary trees used for? They – or, more specifically, a special kind of binary tree known as a *binary search tree* – are convenient structures for manipulating linearly ordered data. There are very efficient algorithms for working with them – *searching* for and *revising* data. There are also good algorithms for *converting* lists into binary search trees and back from them to lists. Moreover, any finite tree can be reconfigured as a binary search tree. We explain these points in turn.

We begin with the concept of a binary search tree. Suppose we have a binary tree and a relation $<$ that linearly orders its nodes. For example, the nodes might be numbers and $<$ the usual numerical order; or the nodes could be expressions of a natural or formal language with $<$ a lexicographic order. Then the binary tree is called *a binary search tree* (modulo that relation) iff each node x is greater than every node in the left subtree of x, and is less than every node in the right subtree of x.

Alice box: Binary search trees

Alice: One moment: does the relation $<$ order the *nodes* themselves, or *labels* on the nodes?

Hatter: As we are defining the notion, it orders the nodes. You could, if you like, treat it as a relation over labels, but it would follow from the definition of a binary search tree that the labelling function must be injective. So we might as well simplify life and think of $<$ as ordering the nodes themselves.

EXERCISE 7.5.1 (WITH SOLUTION)

Which of the four binary trees in Figure 7.6 below are binary search trees, where the order $<$ is the usual order between integers?

Solution: Tree (a) is not a binary search tree for three reasons: $2 < 4$, $10 > 6$, $10 > 9$. Tree (b) isn't either, because $5 > 3$. Nor is (c) a binary search tree, since $7 = 7$. Only (d) is a binary search tree.

Remark: Note that in the definition of a binary search tree, we don't merely require that each node is greater than its left child and less than its right child. We require much more: each node must be greater than *all nodes* in its left subtree, and less than *all nodes* in its right subtree.

Suppose we are given a binary search tree T. How can we *search* it to find out whether it contains an item x? If the tree is empty, then evidently it does not contain x. If the tree is not empty, the following recursive procedure will do the job.

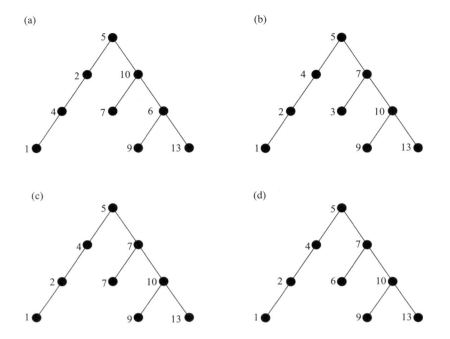

Figure 7.6 *Which are binary search trees?*

Compare x with the root a of T.

If $x = a$, then print *yes*.

If $x < a$ then

    If a has no left child, then print *no*.

    If a does have a left child, apply the recursion to that child.

If $x > a$ then

    If a has no right child, then print *no*.

    If a does have a right child, apply the recursion to that child.

EXERCISE 7.5.2 (WITH SOLUTION)

(a) Execute the algorithm step by step to search for 6 in the last tree of Figure 7.6.

(b) Do the same to search for 8.

(c) For readers familiar with the use of pseudocode: express the above algorithm using pseudocode.

Solution:

(a) The tree is not empty, so we compare 6 with the root 5. Since $6 > 5$ we compare 6 with the right child 7 of 5. Since $6 < 7$ we compare 6 with the left child 6 of 7. Since $6 = 6$ we print *yes*.

(b) The tree is not empty, so we compare 8 with the root 5. Since $8 > 5$ we compare 8 with the right child 7 of 5. Since $8 > 7$ we compare 8 with the right child 10 of 7. Since $8 < 10$ we compare 8 with the left child 9 of 10. Since $8 < 9$ and 9 has no left child, we print *no*.

(c) An elegant way of doing it is as follows:

Search(tree)
begin
   if tree is empty then
      search := no
   else
     if item = root then
       search := yes
     else
      if item < root then
        search := search(left-subtree)
      else
        search := search(right-subtree)
end

The interesting thing about this algorithm is its efficiency. The tree will never be higher than the length of the corresponding list of nodes, and often very much less. At one end of the spectrum, when no node has two children, the node-height of the tree and the length of the list will be the same. At the other end of the spectrum, if every node other than the leaves has two children and all branches are the same length, then the tree will have $2^h - 1$ nodes where h is the node-height of the tree. In other words, the tree will have node-height only $\log_2(n+1)$ where n is the number of its nodes. A search using the algorithm down a branch of length h will thus be very much quicker (in both worst and average cases) than one plodding through a list of all $2^h - 1$ nodes.

There are also algorithms to insert nodes into a binary search tree and to delete nodes from it, in each case ensuring that the resulting structure is still a binary search tree. The algorithm for insertion is quite simple, as it always inserts the new node as a leaf. That for deletion is rather more complex, as it deletes from anywhere in the tree, leaves or interior nodes. We sketch the insertion algorithm, omitting the deletion one.

To *insert* an item x, we begin by searching for it using the search algorithm. If we find x, we do nothing, since it is already there. If we don't find it, we go to the node y where we learned that x is absent from the tree (it should have been in a left resp. right subtree of y, but y has no left resp. right child). We add x as a child of y with an appropriate left or right label.

EXERCISE 7.5.3 (WITH SOLUTION)

Consider again the binary search tree in Figure 7.6. What is the tree that results from inserting the node 8 by means of the algorithm sketched above?

Solution: Searching for 8 in the tree, we reach node 9 and note that 8 should be in its left subtree, but that 9 is a leaf. So we add 8 as left child of 9.

How do we *construct our binary search trees* in the first place? Essentially, by repeated insertion beginning from the empty tree. Suppose that our set of items is given in the form of a list of nine words, e.g. the list $l = $ (this, is, how, to, construct, a, binary, search, tree) and that our ordering relation $<$ for the tree construction is the lexicographic one (i.e. dictionary order). We begin with the empty tree, and search in it for the first item in the list, the word 'this'. Evidently it is not there, so we put it as root. We then take the next item in the list, the word 'is' and search for it in the tree so far constructed. We don't find it, but note that 'is' $<$ 'this' and so we add it as a left child of 'this'. We continue in this way until we have completed the tree.

EXERCISE 7.5.4

(a) Complete the construction of the binary search tree from the word-list $l = $ (this, is, how, to, construct, a, binary, search, tree).

(b) For readers familiar with the use of pseudocode: express the above algorithm for constructing a binary search tree, using pseudocode.

Evidently, when constructed by this algorithm, the shape of the tree depends on the order in which the items are listed. For example, if in the above example the same words were presented in the list $l' = $ (a, binary, construct, how, is, search, this, to, tree), the binary search tree would be a chain going down diagonally to the right, and just as high as the list is long. Clearly, we would like to keep the height of the tree as low as possible. The optimum is to have a complete binary tree (i.e. two children for every node other than the leaves) with all branches of the same length. This is possible only when the list to be encoded has node-length $2^h - 1$ for some $h \geq 0$. Nevertheless, for any value of $h \geq 0$ it is possible to construct a complete binary search tree in which no branch is more than one node longer than any other, and where the longer branches are all to the left of the shorter ones. Moreover, there are good algorithms for carrying out this construction, although we will not describe them in this text.

We end this section by describing one way in which any (finite) ordered tree may be transformed into a binary tree. Let T be any finite ordered tree. To construct our binary tree T' we use exactly the same nodes, and begin the construction by keeping the same root. Let a be any node of the binary tree T' as so far constructed. Its left child will be the first *child* of a in the original tree T, and its right child will be the next *sibling* of a in the original tree T.

EXERCISE 7.5.5

(a) Transform the tree of Figure 7.1 (considered as ordered with left to right ordering of siblings) into a binary tree.

(b) Is it complete?

(c) Give a simple example to show that this transformation, applied to a binary tree, will not in general be the identity.

7.6 Unrooted Trees

We now go back to the rooted tree in Figure 7.1 at the beginning of this chapter, and play around with it. Imagine that instead of the figure made of dots and lines on paper, we have physical model, made of beads connected with pieces of string or wire. We can then pick up the model, push the root a down, and bring up b, say, so that it becomes the root. Or we can rearrange the beads on a table so that the model loses its tree-like shape, and looks more like a road map in which none of the nodes seems to have a special place. When we do this, we are treating the model as an unrooted tree.

7.6.1 Definition of Unrooted Tree

Formally, an *unrooted tree* (alias *undirected tree*) may be defined as any structure (T,S) that can be formed out of a rooted tree (T,R) by taking S to be the symmetric closure of R.

Recall the definition of the symmetric closure S of R: it is the least symmetric relation (i.e. intersection of all symmetric relations) that includes R. Equivalently, $S = R \cup R^{-1}$. Diagrammatically, it is the relation formed by deleting the arrow-heads from the diagram of R (if there were any) and omitting any convention for reading a direction into the links.

EXERCISE 7.6.1

(a) Take the rooted tree in Figure 7.1 and redraw it as an unrooted tree with the nodes scattered haphazardly on the page.

(b) Re-root the unrooted tree that you obtained, by drawing it with node m as root.

(c) Check the equivalence of the two definitions of symmetric closure.

In the context of unrooted trees, terminology changes. Nodes are usually called *vertices*. More important, as well as speaking of links, which are *ordered* pairs (x,y), i.e. elements of the relation, we also need to speak of *edges* (in some texts, *arcs*), identifying them with the *unordered* pairs $\{x,y\}$ such that both (x,y) and $(y,x) \in S$.

It appears that the mathematical concept of an unrooted tree was first articulated in the context of chemistry, by the mathematician Arthur Cayley in 1857. Graphs were already being used to represent the structure of molecules, with vertices (nodes) representing atoms and (undirected) edges representing bonds. Cayley noticed that the saturated hydrocarbons – i.e. the isomers of compounds of the form C_nH_{2n+2} – have a special structure: they are all what we call unrooted trees.

Alice Box: Unrooted trees

Alice: That's a nice, simple definition – provided we are coming to the subject via rooted trees. But what if I wanted to study unrooted trees before the rooted ones? Could I define them in a direct way?

Hatter: No problem. We will do that shortly, after noting some properties ensuing from our present definition.

7.6.2 Properties of Unrooted Trees

Let (T,S) be an unrooted tree, with S the symmetric closure of the link relation R of a rooted tree (T,R). Clearly, S is *connected* (over T), i.e. for any two distinct elements x, y of T, there is an S-path from x to y, i.e. a finite sequence $a_0,\ldots,a_n$ $(n \geq 1)$ of elements of T such that $x = a_0$, $y = a_n$, and each pair $(a_i, a_{i+1}) \in S$. The proof is very straightforward. Since (T,R) is a rooted tree, it has a root a. We distinguish three cases. In the case that $a = x$, by the definition of a rooted tree we have a R-path from x to y, and this is an S-path. In the case that $a = y$, we have a R-path from y to x and thus, running it in reverse (which is legitimate since S is symmetric), we have a S-path from x to y. Finally, if $a \neq x$ and $a \neq y$ we know that there are R-paths from a to each of x, y considered separately. Reversing the path from a to x, we have an S-path from x to a; and continuing it from a to y (i.e. take the composition of the two paths), we end up with one from x to y.

EXERCISE 7.6.2

   Draw tree diagrams to illustrate the three cases of this proof and the constructions made.

   Less obvious is the fact that S is a *minimal* connected relation over T. In other words, no proper subrelation of S is connected over T – for any pair $(x,y) \in S$, the relation $S' = S \setminus \{(x,y)\}$ is *not* connected over T. For suppose that $(x,y) \in S$. Then either $(x,y) \in R$ or $(y,x) \in R$. Consider the former case; the latter is similar. We claim that there is no S'-path from a to y where a is the root of the rooted tree (T,R). Suppose for *reductio ad absurdum* that there is such a S'-path, i.e. a finite sequence $a_0,\ldots,a_n$ $(n \geq 1)$ of elements of T such that $a_0 = a$, $a_n = y$, and each pair $(a_i, a_{i+1}) \in S'$. We may assume without loss of generality that this is a shortest such path, so that in particular never $a_i = a_{i+2}$. Since $(a_{n-1}, a_n) \in S'$, we have either $(a_{n-1}, a_n) \in R$ or conversely $(a_n, a_{n-1}) \in R$. But the former is impossible. Reason: since $a_n = y$ and no node can have two R-parents, we must have $a_{n-1} = x$, so that (x,y) is the last link in the S'-path, contradicting the fact that $(x,y) \notin S'$. Thus the latter alternative $(a_n, a_{n-1}) \in R$ must hold. But then by induction from n to 1, we must have each $(a_{i+1}, a_i) \in R$, for otherwise we would have an a_i with two distinct R-parents. This gives us a R-path from y to the root a, which is impossible by the definition of a rooted tree.

EXERCISE 7.6.3

   (a) Draw a circular diagram to illustrate the above argument.

   (b) In the proof we said: 'We may assume without loss of generality that this is a shortest such path'. Explain why.

The second key property of unrooted trees is that they have no simple cycles. Recall that a *cycle* is a path whose first and last items are the same. So a cycle of an unrooted tree (T,S) is a sequence $a_0,\ldots,a_n$ $(n \geq 1)$ of elements of T with each $(a_i,a_{i+1}) \in S$ and $a_n = a_0$. A *simple cycle* is one with no repeated edges, i.e. for no $i < j < n$ do we have $\{a_i,a_{i+1}\} = \{a_j,a_{j+1}\}$. Expressed in terms of the relation R of the underlying rooted tree: the sequence $a_0,\ldots,a_n = a_0$ never repeats or reverses an R-link. It may however repeat vertices.

For example, the cycles a,b,c,b,a and a,b,c,e,c,b,a are not simple, since each contains both of (b,c) and (c,b). The cycle a,b,c,e,b,c,a is not simple either, as it repeats the link (b,c). On the other hand, the cycle a,b,c,d,c,a is simple, despite the repetition of vertex c.

EXERCISE 7.6.4

(a) Take the first-mentioned cycle a,b,c,b,a and add c in second place, i.e. after the initial a. Is it simple?

(b) Take the last mentioned cycle a,b,c,d,c,a and drop the node b. Is it simple?

Clearly, any unrooted tree with more than one vertex is full of cycles: whenever $(x,y) \in S$ then by symmetry $(y,x) \in S$ giving us the cycle x,y,x. The interesting point is that *it never has any simple cycles*. To see this, suppose for *reductio ad absurdum* that it does have a simple S-cycle $a_0,\ldots,a_n = a_0$. We may assume without loss of generality that this is a shortest one, so that in particular $a_i \neq a_j$ for distinct i,j except for the end points a_0, a_n. We distinguish three cases and find a contradiction in each. *Case 1.* Suppose $(a_0,a_1) \in R$. Then for all i with $0 \leq i < n$ we have $(a_i,a_{i+1}) \in R$, for otherwise some a_{i+1} would have two R-parents a_i and a_{i+2}, which is impossible. Thus the S-cycle $a_0,\ldots,a_n = a_0$ is in fact an R-cycle, which we know from earlier in this chapter is impossible for the link relation R of a rooted tree. *Case 2.* Suppose $(a_n,a_{n-1}) \in R$. A similar argument shows that this case is also impossible. *Case 3.* Neither of the first two cases holds. Then $(a_1 a_0) \in R$ and $(a_{n-1},a_n) \in R$. Since $a_0 = a_n$ and no node of a rooted tree can have two distinct parents, this implies that $a_1 = a_{n-1}$. But that gives us a shorter S-cycle $a_1,\ldots,a_{n-1} = a_1$ which must also be simple, again giving us a contradiction.

EXERCISE 7.6.5

Draw circular diagrams to illustrate the three cases of this proof.

Indeed, we can go further: when (T,S) is an unrooted tree then S is *maximally* without simple cycles. What does this mean? Let (T,S) be an unrooted tree, derived from a rooted tree (T,R). Let x,y be distinct elements of T with $(x,y) \notin S$. Then the structure (T,S') where $S' = S \cup \{(x,y), (y,x)\}$ contains a simple cycle. We will not give a full proof of this, just sketching the underlying construction. In the principal case that x,y are both distinct from the root a of (T,R), then there are unique R-paths from a to each of them. Take the last node b that is common to these two R-paths, and form an S-path from x up to b (using R^{-1}) and then down to y (using R). With (y,x) also available in S', we thus get an S'-cycle from x to x, and it is not difficult to check that this cycle must be simple.

EXERCISE 7.6.6

(a) Draw a tree diagram to illustrate the above proof sketch.

(b) Fill in the details of the above proof by (i) covering the cases that either x or y is the root and (ii) in the case that neither is the root, showing that the constructed cycle is simple, as claimed in the text.

Summarizing what we have done so far in this section, we see that unrooted trees (T,S) satisfy the following conditions:

- S is the symmetric closure of the link relation of some rooted tree (T,R) (by definition)

- S is symmetric, T is connected by S, and S has no simple cycles

- S is symmetric, and T is minimally connected by S

- S is symmetric and is maximally without simple cycles.

In fact, it turns out that these four conditions are mutually equivalent, for any set T and relation $S \subseteq T^2$, so that any one of them could serve as the definition of a rootless tree. The first bulleted condition defined rootless trees out of rooted ones; the remaining ones could be used to define them directly. A relation S that satisfies any one of these conditions, for a given set T, is also known as a *spanning tree* for T.

Alice Box: Explicit versus recursive definitions of unrooted trees

Alice: Thanks, this answers my last query. But one question leads to another. Could we also define unrooted trees recursively, as we did for rooted ones?

Hatter: Of course. Indeed, we need only take the recursive definitions for rooted trees and tweak them a little. Do it as an exercise.

EXERCISE 7.6.7 (WITH PARTIAL SOLUTION)

(a) Without proof, using only your intuition, do the exercise for Alice.

(b) Show that an unrooted tree with n vertices must have $n-1$ edges.

(c) Give an example to show that this numerical relation between vertices and edges is not, however, sufficient to guarantee that a structure (T,S) with S irreflexive and symmetric, is an unrooted tree.

Solution to (b) and (c):

(b) Let (T,S) be an unrooted tree, obtained from a rooted tree (T,R). In an exercise in an earlier section of this chapter, we showed that if T has n elements, then R must have $n-1$ links. But by definition, (T,R) and (T,S) have the same vertices, and the number of edges of (T,S) is the same as the number of links of (T,R). The desired result follows immediately.

(c) The simplest example has four vertices, with three connected in a simple cycle and the fourth not connected to anything.

7.6.3 Finding Spanning Trees

The equivalence of the four bulleted conditions above provides us with a very useful tool. For example, suppose we are given a set A connected by a symmetric and irreflexive relation S over it. There may be many S-paths between vertices, and many simple cycles. In the limit, every element of A may be related to every other one, giving us $n(n-1)/2$ edges, where n is the number of elements of A. That is a lot of information, most of which may be redundant for specific purposes. Often, we need only just enough information to know that A is connected. So the question arises: Is there an algorithm which, given a set A connected by a symmetric and irreflexive relation S over it, finds a minimal symmetric relation $S' \subseteq S$ that connects A? In other words, an algorithm to find a spanning tree for A?

Clearly there is a 'top down' procedure that does the job. We take the given relation S connecting A, and take off edges one by one in such a way as to leave A connected. When we get to a point where it is no longer possible to delete any edge without de-connecting A (which will happen when we have got down to n–1 edges, where n is the number of vertices), we stop. That leaves us with a *minimal* symmetric relation $S' \subseteq S$ connecting A – which is what we are looking for.

But there is also a 'bottom up' procedure that does the job. We begin with the empty set of edges, and add in edges from S one by one in such a way as never to create a simple cycle. When we get to a point where it is no longer possible to add an edge from S without creating a simple cycle (which will happen when we have got up to n–1 edges, where n is the number of vertices), we stop. That leaves us with a *maximal* relation $S' \subseteq S$ without simple cycles. By the last two of the four equivalent conditions bulleted above, S' will also be a *minimal* symmetric relation connecting A – which is what we are looking for.

In general, the 'bottom up' algorithm is much more efficient than the 'top down' one for finding a spanning tree, since it is less costly computationally to check whether a given relation creates a simple cycle than to check whether a given set is connected by a relation.

EXERCISE 7.6.8

For readers familiar with the use of pseudocode: express each of the algorithms above using pseudocode.

In many problems, we need to go further and consider symmetric relations whose edges have numerical weights attached to them. These weights may represent the distance between vertices, or the time or cost involved in passing from one to the other. In this context, we often want to do more than minimize the set of edges in a relation connecting the domain; we may wish to minimize total cost, i.e. minimize *the sum of their weights*. In the usual terminology, we want to find a *minimal spanning tree* for A.

The 'bottom up' algorithm that we have described for finding a spanning tree can be refined to one that finds a minimal spanning tree. However pursuit of this issue would take us beyond the limits of the present book.

FURTHER EXERCISES

7.1. *Properties of rooted trees*

(a) Show that the relation R of a rooted tree (T,R) is always intransitive, in the sense defined in the chapter on relations. *Hint*: Use the explicit definition of a rooted tree, and argue by using proof by contradiction.

(b) Consider the 'new root' definition of a rooted tree. Show that the tree constructed by an application of the recursion step has height (whether node-height or link-height) one larger than the maximal height (in the same sense) of its immediate subtrees.

(c) Do the same with the 'new' leaf definition of a rooted tree.

(d) In the text we defined the notion of a subtree of a tree, but we did not verify that when so defined, it is always itself a tree. Check it.

7.2. *Definitions of rooted trees*

Show the equivalence of the three definitions of a rooted tree – explicit, recursive by 'new root' and recursive by 'new leaf'. *Remark*: This is a challenging exercise, and its answer will have to consist of at least three parts. Perhaps the simplest strategy is to establish a cycle of implications.

7.3. *Labelled trees*

(a) Construct the syntactic decomposition tree of the arithmetic expression $(8{-}(7{+}x)){+}(y^3{+}(x{-}5))$.

(b) Construct the syntactic decomposition tree of the arithmetic expression $-(-(8{-}(7{+}x)))$.

(c) Explain why the syntactic decomposition tree of any arithmetic expression formed from variables and/or constants by the operations of addition, subtraction, multiplication, division, exponentiation (x^y) and taking arbitrary roots ($^x\!\sqrt{y}$), must be a complete binary tree. What happens if one allows into the syntax the one-place operation of sign-reversal (the minus sign)?

(d) Draw the mirror image of the tree in Figure 7.3, with its labels. Write down the arithmetic expression to which it corresponds, in both standard and Polish notation.

7.4. *Binary search trees*

(a) Using the construction algorithm given in the text, construct a binary search tree for the list of letters (g,c,m,b,i,h,o,a) where the ordering relation $<$ is the usual alphabetical one.

(b) In the tree you have constructed, trace the steps in a search for the letter i, using the search algorithm in the text.

(c) To the tree you have constructed, add the letter f, using the insertion algorithm in the text.

7.5. *Unrooted trees*

Consider the *complete* graph (G,R) with five vertices, i.e. the graph in which there is an edge between each vertex and every other vertex.

(a) Construct a spanning tree for this graph by the top-down method, showing by successive diagrams each step.

(b) Construct a different spanning tree by the bottom-up method, again showing your steps by successive diagrams.

Selected Reading

Almost all introductory texts of discrete mathematics have a chapter on trees. Usually it follows one on graphs – in contrast to our procedure here, which treats trees without getting into the more general theory of graphs. Among the most widely texts used are:

John A. Dossey et al. *Discrete Mathematics*. Pearson, 2006 (fifth edition) Chapters 4 (graphs) and 5 (trees).

Ralph P. Grimaldi *Discrete and Combinatorial Mathematics*. Addison Wesley, 2004 (fifth edition), Chapters 11 (graphs) and 12 (trees).

Richard Johnsonbaugh *Discrete Mathematics*. Pearson, 2005 (sixth edition) Chapters 8 (graphs) and 9 (trees).

Kenneth Rosen *Discrete Mathematics and its Applications*. McGraw Hill, 2007 (sixth edition), Chapters 9 (graphs) and 10 (trees).

8

Yea and Nay: Propositional Logic

Chapter Outline

We have been using logic on every page of this book – in every proof, verification and informal justification. In the first four chapters we inserted some 'logic boxes'; they gave just enough to be able to follow what was being done. Now we gather the material of these boxes together and develop their principles. Logic thus emerges as both a tool for reasoning and an object for study.

We begin by explaining different ways of approaching the subject, and situating the kind of logic that we will be concerned with. We give an outline of the *purely structural part* of logic – that is, the part that is independent of the expressive power and internal features of the language under consideration – and then go on to a more detailed account of *truth-functional*, also known as classical propositional, logic. The basic topics will be the *truth-functional connectives*, the family of *tautologicality concepts*, the availability of *normal forms* and *unique minimalities*, the use of *semantic decomposition trees* as a shortcut method for testing status, and finally a sketch of *natural deduction* with its two components, *enchaining* and *indirect inference*. The following chapter will get into the quantifiers.

Readers are advised to flip back to the logic boxes in the first four chapters to get up to speed for these two.

D. Makinson, *Sets, Logic and Maths for Computing*,
DOI: 10.1007/978-1-84628-845-6_8, © Springer-Verlag London Limited 2008

8.1 What is Logic?

What we will be studying in these chapters is only one part of logic, but a very basic part. In its most general form, logic is best seen as the study of *reasoning*, to use an old-fashioned term, *belief management* for a more fashionable one. It concerns ways in which agents (human or other) may develop and shape their beliefs, by inference, organization, and change.

- *Inference* is the process by which one proposition is accepted on the basis of others, as justified by them.

- *Organization* is the business of getting whatever we believe into an easily exploitable pattern. It is particularly important in mathematics, where it can take the form of *axiomatization*. Even when axiomatization is not available or appropriate, some kind of conceptual structuring is still needed to find a global view.

- *Change* takes place when we decide to abandon a belief – a process known to logicians as *contraction*. It can also take the form of *revision*, where we accept something that we previously ignored or rejected, at the same time making sufficient contractions to maintain consistency of the whole. A closely related form of belief change, of particular interest to computer science, is *update*. This is the process of modifying our stock of beliefs about a domain to keep up with changes that are taking place in the world. Evidently, it is very close to revision in the sense that we have described, but there are also subtle differences.

Of all these processes of belief management, the most basic is *inference*. It reappears in all of the others, just as sets reappear in relations, functions, probability and trees. For this reason, introductory logic books restrict themselves to inference, leaving other concerns for advanced work. Even within that sphere, they look at only one part of the subject, though admittedly the most fundamental one – *deductive* as contrasted with *non-deductive* inference. We must do the same.

That this is a real limitation becomes apparent if you reflect on the kind of reasoning that you carry out in daily life, outside the study and without the aid of pen and paper. Even when it takes the form of inference, it is seldom fully deductive. The conclusions you reach are (hopefully) plausible, reasonable, probable, or convincing given the assumptions made; but they are rarely if ever absolutely certain given those same assumptions. There is the possibility, perhaps remote, that *even if* the assumptions are true, the conclusion *might* still be false.

But within mathematics, the game is quite different. There we make systematic use of fully deductive inference, rendering the conclusions certain given the assumptions made. This is not to say that there is any certainty in the assumptions themselves, nor for that matter in the conclusions. It lies in the *link* between them: it is impossible for the premises to be true without the conclusion also being so.

That is the kind of reasoning we will be studying in these chapters. It provides a basis that is needed before trying to tackle any other kind of inference or belief management. Its study goes back two thousand years; in it modern form, it began to take shape in the middle of the nineteenth century.

8.2 Structural Features of Consequence

Inferences are made up of *propositions*, alias *assertions* or *statements*. As in the logic boxes of earlier chapters, we will use lower-case Greek letters α, β, ... for them. Shortly we will need to look at their internal structure, but for the moment we treat them as unanalysed objects. All we need to assume in this section is that propositions may sometimes be true (unfortunately, not always so), and that that we would like our inferences to preserve truth. Deductive inferences should guarantee that we *never lose it*.

An inference links a set A of propositions, called its *premises*, and a proposition β called its conclusion. Typically A is finite, so that $A = \{\alpha_1, \ldots, \alpha_n\}$ for some natural number $n \geq 0$, and we will attend mainly to that case. But for smooth formulation it will sometime be convenient to cover infinite sets of premises as well. It may seem odd that we are allowing the case that $n = 0$, i.e. that A may be the empty set, but that will become clear as we proceed.

A proposition β is said to be a *logical consequence* (briefly, when context is clear: *consequence*) of a set A of propositions, and we write $A \vdash \beta$, iff it is impossible for its conclusion to fail to be true while its premises are all true. Looking at the relation conversely, we also say that A *logically implies* β (briefly: *implies*) in this situation. Note that this does not require the premises to be true, nor the conclusion to be so; merely that the former is impossible without the latter. The sign $\vdash$ is called the *gate* or *turnstile* symbol. To reduce notation, it is customary to drop parentheses and write $\alpha \vdash \beta$ instead of $\{\alpha\} \vdash \beta$, often even $\alpha_1, \ldots, \alpha_n \vdash \beta$ in place of $\{\alpha_1, \ldots, \alpha_n\} \vdash \beta$.

What are the general properties of the consequence relation? Perhaps the most basic is the property known as *identity*: every proposition is a consequence of any set of which it is an element. That is,

$$A \vdash \beta \text{ whenever } \beta \in A.$$

The justification is immediate. Suppose $\beta \in A$. Then it is impossible for β to fail to be true when every element of A (thus including β itself) is so. Taking the particular case where A is a singleton $\{\alpha\}$, the principle also tells us that $\alpha \mid\!- \alpha$ for any formula α.

Alice Box: Using logic in order to understand logic

Alice: This is weird. You are using logic to explain logic!

Hatter: Indeed we are. And there is no way around it. The extraordinary thing is that it can be done. It is rather like using your eyes to study, in a magnifying mirror, the structure of your eyes.

Alice: The definition of consequence made use of the notion of *truth*. But you did not mention *falsehood*, only the presence or absence of truth. Any reason for this?

Hatter: We will need the notion of falsehood, with the principle of *bivalence* linking truth and falsehood, in the next section when we begin analysing the internal structure of propositions. We could have used it already here, but we don't really need it yet.

A second property of the consequence relation, less trivial, is known as *cumulative transitivity* (alias *cut*): whenever β_1, β_2, ... are all consequences of a set A, and γ is a consequence of A taken together with β_1, β_2, ..., then γ is a consequence of A itself. That is:

Whenever $A \mid\!- \beta$ for all propositions $\beta \in B$, and $A \cup B \mid\!- \gamma$ then $A \mid\!- \gamma$.

Justification: Suppose all propositions in A are true. Then since $A \mid\!- \beta$ for all propositions $\beta \in B$, each such β must be true, so that all propositions in $A \cup B$ are true. Hence since $A \cup B \mid\!- \gamma$ we know that γ is true.

Another way of putting this: we may accumulate validly obtained conclusions into our initial premises without ever being led astray. When B is a singleton $\{\beta\}$, this gives us the following particular case, rather easier to grasp:

Whenever $A \mid\!- \beta$ and $A \cup \{\beta\} \mid\!- \gamma$ then $A \mid\!- \gamma$

When A is also a singleton $\{\alpha\}$ this comes down to:

Whenever $\alpha \mid\!- \beta$ and $\{\alpha, \beta\} \mid\!- \gamma$ then $\alpha \mid\!- \gamma$

Innocent as cumulative transitivity may be for deductive reasoning, it is not in general appropriate for all forms of non-deductive inference. In particular, probabilistic inference does not satisfy it! But that is beyond our present concerns.

A third property is known as *monotony* (or monotonicity). It tells us that whenever γ is a consequence of a set A, then it remains a consequence of any set obtained by adding any other propositions to A. In other words, increasing the stock of premises can never lead us to drop a conclusion:

$$\text{Whenever } A \vdash \gamma \text{ and } A \subseteq B \text{ then } B \vdash \gamma$$

Not that this is for arbitrary choice of B – even when B is inconsistent with A, or with γ. While monotony holds for deductive inference, again it is quite unacceptable for non-deductive reasoning, whether probabilistic or expressed in qualitative terms. That is why some recently developed systems for uncertain reasoning have been dubbed 'nonmonotonic logics'.

Taken together, these three principles are often known as the *Tarski conditions* on logical consequence as a relation, named after the Polish logician Alfred Tarski.

EXERCISE 8.2.1 (WITH PARTIAL SOLUTION)

(a) Justify the principle of monotony for deductive inference $\vdash$ in the same way as we justified the principles of identity and cumulative transitivity.

(b) Show that monotony may equivalently be expressed thus: whenever $A \vdash \gamma$ then $A \cup B \vdash \gamma$.

(c) Show that monotony may also be expressed as: whenever $A \cap B \vdash \gamma$ then $A \vdash \gamma$.

(d) Sketch a simple intuitive example of non-deductive reasoning (say, concerning the guilt of a suspect in a criminal case) that illustrates how the principle of monotony may fail in such contexts.

(e) Obtain the plain transitivity of the relation $\vdash$ of logical consequence from its cumulative transitivity and monotony.

(f) Is plain transitivity always appropriate for non-deductive inference? Give an intuitive counterexample or explanation.

Solution to (b) and (f):

(b) Summary proof: Clear from the fact that $A \subseteq B$ iff $B = A \cup B$. Detailed verification: Consider the two directions separately. In one direction: assume the version in the text and suppose $A \vdash \gamma$; we want to show that $A \cup B \vdash \gamma$. But $A \subseteq A \cup B$ and so $A \cup B \vdash \gamma$. In the other direction: assume the version in the exercise and suppose both $A \mathrel{|\!\!-} \gamma$ and $A \subseteq B$; we want to show that $B \mathrel{|\!\!-} \gamma$. But since $A \subseteq B$ we have $B = A \cup B$, so $B \mathrel{|\!\!-} \gamma$.

(f) No, plain transitivity fails for non-deductive inference. One way of illustrating this is in terms of an intuitive notion of risk. When we pass from α to β non-deductively, there is a small risk of losing truth, and passage from β to γ may likewise incur a small risk. Taking both steps compounds the risk, which may thereby exceed a reasonable threshold that each of the separate ones respects. In effect, the strength of the chain may be *even less than* the strength of its weakest link. *Remark.* This informal explanation is all we need here. It can be made precise in terms of probabilities and qualitative analyses, but that is beyond the scope of this book.

We mention a last structural condition, which links the finite with the infinite and is usually known as *compactness*. It says that whenever β is a consequence of a set A of propositions, then it is a consequence of some finite subset of A. In other words:

Whenever $A \mathrel{|\!\!-} \beta$ then there is a finite subset $A' \subseteq A$ with $A' \mathrel{|\!\!-} \beta$.

Note carefully the word *some* here. The principle does not tell us *which* subsets A' of A are strong enough to do the job; it only tells us that there is at least one, without letting us know what it is, or even how big it might be. Moreover the choice and size of A' will in general depend on the conclusion β under consideration, as well as on A. Given $A \mathrel{|\!\!-} \beta$ and $A \mathrel{|\!\!-} \gamma$ compactness tells us that there are finite sets $A_1, A_2 \subseteq A$ with $A_1 \mathrel{|\!\!-} \beta$ and $A_2 \mathrel{|\!\!-} \gamma$; but we will not in general have $A_1 \mathrel{|\!\!-} \gamma$ or $A_2 \mathrel{|\!\!-} \beta$.

The justification of compactness is rather more subtle than that of identity, cumulative transitivity or monotony; indeed it requires a closer analysis of the internal structure of propositions. It holds for most (though not all) forms of deductive inference. As this book focuses on the finite case, we will not follow these matters further, just leaving the reader with the following easy exercise.

EXERCISE 8.2.2

Use compactness and monotony to show that whenever A is an infinite set of propositions and $A \mathrel{|\!\!-} \beta$ then there are infinitely many finite subsets $A' \subseteq A$ with $A' \mathrel{|\!\!-} \beta$.

Logicians have found it convenient to express logical consequence in an equivalent manner, as a *function* (alias operation) rather than as a *relation*. Let A be any set of propositions, and define $Cn(A) = \{\beta: A \vdash \beta\}$. The function Cn thus takes sets of propositions to sets of propositions; we can write it as $Cn: \mathrm{P}(L) \to \mathcal{P}(L)$, where L is the set of all propositions of the language under consideration and $\mathcal{P}$ is the powerset operation, familiar from the chapter on sets. The name Cn is chosen to recall the word 'consequence'. So formulated, the operation has the following properties:

Inclusion : $A \subseteq Cn(A)$

Idempotence : $Cn(A) = Cn(Cn(A))$

Monotony :   Whenever $A \subseteq B$ then $Cn(A) \subseteq Cn(B)$.

These reflect the three basic properties of consequence as a relation, and can be derived from them. Taken together, they are also called *the Tarski conditions* for logical consequence as an operation. The term *closure operation* is also often used for any function satisfying them. Evidently, the operational formulation is much more succinct than the relational one, and for this reason it is very useful when one gets further into the subject; but it does take a bit of getting used to. In this book we will usually talk in terms of a relation.

EXERCISE 8.2.3

(a) Show that Tarski condition of inclusion, for logical consequence as an operation, follows from identity for consequence as a relation, via the definition $Cn(A) = \{\beta: A \vdash \beta\}$ given in the text.

(b) What would be the natural way of defining consequence as a relation from consequence as an operation?

(c) For *aficionados* of the infinite: Devise a succinct expression of compactness in the language of consequence as an operation. *Hint*: Try filling in the dots in $Cn(A) = \cup\{\dots\}$.

(d) Given the relation $\vdash$ of logical consequence, what would be the natural definition of logical equivalence between two propositions? Check that your candidate relation is reflexive, symmetric and (especially) transitive.

(e) In the same vein, what would be the natural definition of logical equivalence between two *sets* of propositions?

8.3 Truth-Functional Connectives

Having looked at the general concept of deductive inference and articulated its basic properties as a relation and as an operation, we now peer into the interior of propositions to examine some of the ways in which they may be built up. We can then see how the general definition of logical consequence, given in the preceding section, manifests itself in this context.

The simplest ways of building new propositions out of old are by means of the truth-functional connectives 'not', 'and', 'or', 'if', 'iff', etc. We have already introduced these one by one in logic boxes in the first four chapters, and we will not repeat everything here. If you have not already done it, you should revise those boxes before going further. For easy reference, however, we recall the truth-tables. The one-place connective 'not' has the table:

Table 8.1 Truth-table for negation.

α	$\neg\alpha$
1	0
0	1

The two-place connectives 'and', 'or', 'if', 'iff' have the tables, grouped together:

Table 8.2 Truth-table for familiar two-place connectives.

α	β	$\alpha\wedge\beta$	$\alpha\vee\beta$	$\alpha\rightarrow\beta$	$\alpha\leftrightarrow\beta$
1	1	1	1	1	1
1	0	0	1	0	0
0	1	0	1	1	0
0	0	0	0	1	1

Here 1 is for truth, and 0 is for falsehood. We are thus using a little more machinery than in the section on consequence relations, where truth alone was used. A basic assumption made in these tables is the *principle of bivalence*: every proposition is either true, or false, but not both. In other words, the truth-values of propositions may be represented by a function with domain the set of all propositions of the language under consideration (about which more shortly) into the two-element set $\{1,0\}$.

Alice Box: The principle of bivalence

Alice: What happens if we relax the principle of bivalence?

Hatter: There are two main ways of going about it. One is to allow that there may be truth-values other than truth and falsehood. This gives rise to the study of what is called *many-valued logic*.

(Continued)

> *Alice Box:* (Continued)
>
> *Alice*: And the other?
>
> *Hatter*: We can allow that the values may not be exclusive, so that a proposition may be both true and false. That can also be accommodated within many-valued logic by using new values to represent subsets of the old values. For example, we might use four values (the two old ones and two new ones) to represent the four subsets of the classical values 1 and 0. There are also systems of logic that avoid truth values altogether. But these are all *non-classical*, and beyond our compass. In any case, classical logic remains the standard base.

It is time to look more closely at the concept of a *truth-function* in two-valued logic. A one-place truth-function is simply a function on domain $\{1,0\}$ into $\{1,0\}$. A two-place one is a function on $\{1,0\}^2$ into $\{1,0\}$, and so on. This immediately suggests all kinds of questions. How many one-place truth-functions are there? Two-place, and generally n-place? Can every truth-function be represented by a truth-table? Are the specific truth functions given in the tables above sufficient to represent all of them, or should we introduce further logical connectives to do so?

Clearly, there are just four one-place truth-functions. Each can be represented by a truth-table. In the leftmost column we write the two truth values 1,0. In the remaining columns we write the possible values of the functions for those two values of their arguments:

Table 8.3 The one-place truth-functions.

α	$f_1(\alpha)$	$f_2(\alpha)$	$f_3(\alpha)$	$f_4(\alpha)$
1	1	1	0	0
0	1	0	1	0

EXERCISE 8.3.1 (WITH SOLUTION)

(a) Which of these four truth-functions f_i corresponds to negation?

(b) Can you express the other three truth-functions in terms of connectives with which you are familiar ($\neg, \wedge, \vee, \rightarrow, \leftrightarrow$)?

Solution:

(a) Obviously, f_3.

(b) f_2 is the identity function, i.e. $f_2(\alpha) = \alpha$, so we don't need to represent it by more than α itself. f_1 is the constant function with value 1, and so can be represented as, say, $\alpha \vee \neg \alpha$ or $\alpha \rightarrow \alpha$ or $\alpha \leftrightarrow \alpha$. Finally, f_4 is the constant function with value 0, and can be represented as $\alpha \wedge \neg \alpha$, or as the negation of any of those for f_1.

Going on to the two-place truth-functions, the Cartesian product $\{1,0\}^2$ evidently has $2^2 = 4$ elements, which may be listed in a table with four rows, as in the tables for the familiar two-place connectives. We always write these four rows in a standard order, as in those tables. For each pair $(a,b) \in \{1,0\}^2$ there are evidently two possible values for the function, which gives us $4^2 = 16$ columns to fill in, i.e. 16 truth-functions.

EXERCISE 8.3.2

(a) Write out a table for all 16 two-place truth-functions. *Hints*: You will need 4 rows for the four ordered pairs $(a,b) \in \{1,0\}^2$, plus a top one for the labels. For easy communication, the rows should be written in the standard order. You will need 16 columns for the sixteen functions, plus 2 for their two arguments. Again, these columns should be given in standard order: begin with the column $(1,1,1,1)$ and proceed in the natural way to the last column $(0,0,0,0)$. The principle for constructing each column from its predecessor: take the last 1 in the preceding column, change it to 0, and replace all 0 s lower in the column to 1.

(b) Draw a binary tree of link-depth 4 in which the 16 truth-functions of two places are represented by branches, with standard order from left to right.

(c) Can you express all sixteen truth-functions in terms of the familiar ones $\neg, \wedge, \vee, \rightarrow, \leftrightarrow$?

In general, for the n-place truth-functions $f(a_1,..,a_n)$ with each $a_i \in \{1,0\}$, we need $r = 2^n$ rows in the table (plus a top one for the labels) and thus 2^r columns for the truth-functions (in addition to n columns on the left for the n arguments). The number of truth-functions is thus doubly exponential in the number of their arguments.

The familiar truth-functions are all two-place ones. Is this a gap in their expressive power, or can every truth-function, of no matter how many places, be captured using them? Fortunately, there is no gap. Every truth-function, of any finite number of places, may be represented using at most the three connectives $\neg, \wedge, \vee$. To see how, we return to the two-place functions. In the last exercise, they were all expressed in terms of $\neg, \wedge, \vee, \rightarrow, \leftrightarrow$, but perhaps in a haphazard manner. Now we do it using just $\neg, \wedge, \vee$, in a systematic fashion. Take an arbitrary two-place truth-function f_i, say f_3, with its table:

Table 8.4 Sample two-place truth-table.

α	β	$f_3(\alpha,\beta)$
1	1	1
1	0	1
0	1	0
0	0	1

Take the rows in which $f_i(\alpha,\beta) = 1$ – there are three of them in this instance. For each such row, form the conjunction $\pm\alpha\wedge\pm\beta$ where $\pm$ is empty or negation according as the argument has 1 or 0. This gives us the three conjunctions $\alpha\wedge\beta$ (for the first row), $\alpha\wedge\neg\beta$ (for the second row), and $\neg\alpha\wedge\neg\beta$ (for the fourth row). Form their disjunction: $(\alpha\wedge\beta)\vee(\alpha\wedge\neg\beta)\vee(\neg\alpha\wedge\neg\beta)$. Then this will express the same truth-function f_3.

Why? By the table for disjunction, it will come out true just when at least one of the three disjuncts is true. But the first disjunct is true in just the first row, the second is true in just the second row, and the third is true in just the last row. So the constructed expression has exactly the same truth-table as f_3, i.e. it expresses that truth-function.

EXERCISE 8.3.4

(a) Of the 16 two-place truth-functions, there is one that does *not* have a full disjunctive normal form of the kind described. Which is it? Show how we can still represent it using $\neg, \wedge, \vee$.

(b) Draw a truth-table for some three-place truth-function, and express it using $\neg, \wedge, \vee$ in the same way.

Thus every truth-function may be expressed using only the connectives $\neg, \wedge, \vee$. Have we reached the limit, or can we do the job with even fewer connectives? We will return to this question after clarifying some fundamental concepts.

8.4 Tautologicality

A special feature of logic, as compared with most other parts of mathematics, is the very careful attention that it gives to the *language* in which it is formulated. Whereas the theory of trees, say, is about certain kinds of abstract structure – arbitrary sets equipped with a relation satisfying a certain condition – logic is about the interconnections between certain kinds of language and structures that may be used to interpret them. A good deal of logic thus *looks at the language* as much as at the structures in which it is interpreted. This can take quite some time to get used to, since ordinarily in mathematics we look *through* the language at the structures alone.

8.4.1 The Language of Propositional Logic

Our language should be able to express all truth-functions. Since we know that the trio $\neg$, $\wedge$, $\vee$ are together sufficient for the task, we may confine ourselves to them. They are the *connectives* of the language. We take some stock of expressions p_1, p_2, $p_3,\ldots$, understood intuitively as representing propositions, and call them *elementary letters* (in some texts: *propositional variables*). This set may be finite or infinite; the usual convention is to take it either as finite but of unspecified cardinality, or as countably infinite. As subscripts are a pain, we usually write elementary letters as $p,q,r,\ldots$

The *formulae* of our language are expressions that can be obtained recursively from elementary letters by applying connectives. If the chapter on recursion and induction has not been forgotten entirely, it should be clear what this means: the set of formulae is the least set L that contains all the elementary letters and is closed under the connectives, i.e. the intersection of all such sets. That is, whenever $\alpha \in L$ then $(\neg\alpha) \in L$, and whenever $\alpha,\beta \in L$ then so is each of $(\alpha\wedge\beta)$ and $(\alpha\vee\beta)$. Only expressions that can be formed in a finite number of such steps are counted as formulae.

In the chapter on trees we have already seen why brackets are needed in formulae. For example, we need to be able to distinguish $((p\wedge q)\vee r)$ from $(p\wedge(q\vee r))$ and likewise $(\neg(p\wedge q))$ from $((\neg p)\wedge q)$. We have also seen how the brackets may in principle be dispensed with if we adopt a prefix (Polish) or postfix (reverse Polish) notation, and have learned how to draw the syntactic decomposition tree of a formula. We have also agreed to make reading easier by omitting brackets when this can be done without ambiguity (e.g. omitting the outermost brackets, which we will always do) or when the ambiguity is resolved by standard grouping conventions (e.g. reading $\neg p\wedge q$ as $(\neg p)\wedge q$ rather than as $\neg(p\wedge q)$.

EXERCISE 8.4.1

For revision, draw the syntactic decomposition trees of the two formulae $(p \wedge q) \vee r$ and $p \wedge (q \vee r)$, and also write each of them in both Polish and in reverse Polish notation. Likewise for the formulae $\neg(p \wedge q)$ and $(\neg p) \wedge q$.

With this out of the way, we can get down to more serious business, defining the fundamental concepts of classical propositional logic.

8.4.2 Assignments and Valuations

An *assignment* is a function $v: E \rightarrow \{1,0\}$ on the set E of elementary letters of the language into the two-element set $\{1,0\}$. Roughly speaking, assignments correspond to left-hand parts of rows of a truth-table. By structural induction on formulae, it follows that for each assignment $v: E \rightarrow \{1,0\}$ there is a unique function $v^+: L \rightarrow \{1,0\}$, where L is the set of all formulae, that agrees with v on E (that is, $v^+(p) = p$ for every elementary letter p) and also satisfies the truth-table conditions, i.e. $v(\neg\alpha) = 1$ iff $v(\alpha) = 0$, $v(\alpha \wedge \beta) = 1$ iff $v(\alpha) = v(\beta) = 1$, and $v(\alpha \vee \beta) = 0$ iff $v(\alpha) = v(\beta) = 0$. Such a v^+ is called a *valuation* of formulae. Following the editorial maxim of minimizing notational fuss, we almost always 'abuse notation' by dropping the superscript from v^+ and writing it too as v.

We are interested in a group of notions that may be called the *tautologicality concepts*. There are four basic ones. Two are *relations* between formulae: tautological implication, tautological equivalence. The other two are *properties* of formulae: those of being a tautology and of being a contradiction. They are intimately linked to each other.

8.4.3 Tautological Implication

We begin with the relation of tautological implication. Let A be a set of formulae, and β an individual formula. We say that A *tautologically implies* β (or: β is a *tautological consequence* of A) and write $A \vdash \beta$ iff there is no valuation v such that $v(\alpha) = 1$ for all $\alpha \in A$ but $v(\beta) = 0$. In other words, for every valuation v, if $v(\alpha) = 1$ for all $\alpha \in A$ then $v(\beta) = 1$. When A is a singleton $\{\alpha\}$, this says: $\alpha \vdash \beta$ iff $v(\beta) = 1$ whenever $v(\alpha) = 1$. Note that we do not require the converse.

In terms of truth-tables for the singleton case $\alpha \vdash \beta$, this means if we draw up a truth-table that covers all the elementary letters that occur in α or in β (for they might not have exactly the same letters in them), then every row which has a 1 under α also has a 1 under β. For the case where the premise set is more than a singleton, the check is the same except that we consider the rows in which there is a 1 under all of the $\alpha \in A$.

Warning: The symbol |- is *not* one of the connectives of propositional logic like $\neg, \wedge, \vee, \rightarrow, \leftrightarrow$. Whereas $p \rightarrow q$ is a formula of the language L of propositional logic, p |- q is not. Rather, |- is a symbol that we use when talking *about* formulae of propositional logic. In handy jargon, we say that it belongs to our *metalanguage* rather than to the *object language*. This distinction takes a little getting used to, but it is very important. Neglect can lead to inextricable confusion.

Here is a table of some of the more important tautological implications, with their usual names (some going back a thousand years). Most have a singleton premise set. When two formulae are separated by a comma, they form a two-element premise set. Thus, for example, the premise set for modus ponens is the pair $\{\alpha, \alpha \rightarrow \beta\}$.

Table 8.5 Some important tautological implications.

Name	LHS	RHS
Simplification	$\alpha \wedge \beta$	α
	$\alpha \wedge \beta$	β
Conjunction	α, β	$\alpha \wedge \beta$
Disjunction	α	$\alpha \vee \beta$
	β	$\alpha \vee \beta$
Modus Ponens	$\alpha, \alpha \rightarrow \beta$	β
Modus Tollens	$\neg \beta, \alpha \rightarrow \beta$	$\neg \alpha$
Disjunctive Syllogism	$\alpha \vee \beta, \neg \alpha$	β
Transitivity	$\alpha \rightarrow \beta, \beta \rightarrow \gamma$	$\alpha \rightarrow \gamma$
Material implication	β	$\alpha \rightarrow \beta$
	$\neg \alpha$	$\alpha \rightarrow \beta$
Limiting cases	$\alpha \wedge \neg \alpha$	any formula
	any formula	$\beta \vee \neg \beta$

We check that modus tollens is a tautological implication, by using a truth-table.

Table 8.6 Verification of modus tollens.

p	q	$p \rightarrow q$	$\neg q$	$\neg p$
1	1	1	0	0
1	0	0	1	0
0	1	1	0	1
0	0	1	1	1

The four assignments are given in the left part of the table. The resulting values of $p \rightarrow q$, $\neg q$, $\neg p$ are calculated step by step, from the inside to the outside (one might also say from the bottom up), i.e. from the leaves of the syntactic decomposition trees for the three formulae to their roots (very short trees in this case).

Is there a row in which the two premises $p \rightarrow q$, $\neg q$ get 1 while the conclusion $\neg p$ gets 0? No, so the premises tautologically imply the conclusion.

Alice Box: Elementary letters vs arbitrary formulae

Alice: Wait a moment! In the list of tautological implications, you used the symbols α, β which are meant to range over arbitrary formulae of your language. So modus ponens says that for all formulae α, β, the premise-set $\{\neg\beta, \alpha \rightarrow \beta\}$ tautologically implies the conclusion $\neg\alpha$. In other words, $\neg\alpha$ is a tautological consequence of $\{\neg\beta, \alpha \rightarrow \beta\}$. But in your truth-table, you showed this only for the special case that α, β are elementary letters p, q. Isn't this a bit too fast?

Hatter: Nice point! Indeed, it was a bit fast. But it is legitimate. This is because of a very important theorem about tautological consequence. Whenever a tautological implication holds, and we make a *substitution* uniformly through premises and conclusion, then the tautological implication still holds. In our example of modus tollens, given that $\{\neg q, p \rightarrow q\}$ tautologically implies $\neg p$ where p, q are elementary letters, we know that for arbitrary formulae α and β, the premise-set $\{\neg\beta, \alpha \rightarrow \beta\}$ tautologically implies $\neg\alpha$.

Alice: Proof?

Hatter: Let's get back to it a bit later in the chapter. We will need to give a recursive definition of what a substitution function is, so that we may then build an inductive proof to ride on it.

EXERCISE 8.4.2

(a) Draw truth-tables to check out each of disjunctive syllogism, transitivity, and the two material implication consequences.

(b) Check the two limiting cases (choosing your own formulae in each case), and also explain in general terms why they hold.

(c) Verify that tautological consequence satisfies the three structural Tarski conditions given earlier in the chapter. *Hint*: Your verifications will be like those given in general terms then, but based specifically on the definition of tautological implication.

Once understood, the entries in Table 8.5 should be committed firmly to memory.

8.4.4 Tautological Equivalence

In a tautological implication such as disjunction, $\alpha \mathrel{|-} \alpha\lor\beta$, while the left implies the right, the converse does not in general hold: $p\lor q \mathrel{|\!\!\!/} p$ when p, q are distinct elementary letters, although $\alpha\lor\beta \mathrel{|-} \alpha$ does of course hold for *some* choices of α, β.

EXERCISE 8.4.3

Show the former and give an example of the latter.

When each of two formulae tautologically implies the other, we say that the two are tautologically equivalent. That is, α is *tautologically equivalent* to β, and we write $\alpha \mathrel{-||-} \beta$, iff both $\alpha \mathrel{|-} \beta$ and $\beta \mathrel{|-} \alpha$. Equivalently: $\alpha \mathrel{-||-} \beta$ iff $v(\alpha) = v(\beta)$ for every valuation v.

In terms of truth-tables: when we draw up a truth-table that covers all the elementary letters that occur in α or in β (again, they might not have exactly the same letters in them), we require that the column for α comes out exactly the same as the column for β.

Once again, the symbol $\mathrel{-||-}$ does not belong to the object language, but is part of our metalanguage.

Here is a table of the most important tautological equivalences that can be expressed in up to three elementary letters. They should also be committed to memory after being understood.

Table 8.7 Some important tautological equivalences.

Name	LHS	RHS
Double negation	α	$\neg\neg\alpha$
Commutation for $\land$	$\alpha\land\beta$	$\beta\land\alpha$
Association for $\land$	$\alpha\land(\beta\land\gamma)$	$(\alpha\land\beta)\land\gamma$
Commutation for $\lor$	$\alpha\lor\beta$	$\beta\lor\alpha$
Association for $\lor$	$\alpha\lor(\beta\lor\gamma)$	$(\alpha\lor\beta)\lor\gamma$
Distribution of $\land$ over $\lor$	$\alpha\land(\beta\lor\gamma)$	$(\alpha\land\beta)\lor(\alpha\land\gamma)$
Distribution of $\lor$ over $\land$	$\alpha\lor(\beta\land\gamma)$	$(\alpha\lor\beta)\land(\alpha\lor\gamma)$
Absorption	α	$\alpha\land(\alpha\lor\beta)$
	α	$\alpha\lor(\alpha\land\beta)$
Expansion	α	$(\alpha\land\beta)\lor(\alpha\land\neg\beta)$
	α	$(\alpha\lor\beta)\land(\alpha\lor\neg\beta)$
De Morgan	$\neg(\alpha\land\beta)$	$\neg\alpha\lor\neg\beta$
	$\neg(\alpha\lor\beta)$	$\neg\alpha\land\neg\beta$
	$\alpha\land\beta$	$\neg(\neg\alpha\lor\neg\beta)$
	$\alpha\lor\beta$	$\neg(\neg\alpha\land\neg\beta)$
Contraposition	$\alpha\to\beta$	$\neg\beta\to\neg\alpha$
	$\alpha\to\neg\beta$	$\beta\to\neg\alpha$
	$\neg\alpha\to\beta$	$\neg\beta\to\alpha$

Table 8.7 (continued)

Name	LHS	RHS
Import/export	$\alpha\rightarrow(\beta\rightarrow\gamma)$	$(\alpha\wedge\beta)\rightarrow\gamma$
	$\alpha\rightarrow(\beta\rightarrow\gamma)$	$\beta\rightarrow(\alpha\rightarrow\gamma)$
Consequentia mirabilis	$\alpha\rightarrow\neg\alpha$	$\neg\alpha$
(miraculous consequence)	$\neg\alpha\rightarrow\alpha$	α
Commutation for $\leftrightarrow$	$\alpha\leftrightarrow\beta$	$\beta\leftrightarrow\alpha$
Association for $\leftrightarrow$	$\alpha\leftrightarrow(\beta\leftrightarrow\gamma)$	$(\alpha\leftrightarrow\beta)\leftrightarrow\gamma$
$\neg$ through $\leftrightarrow$	$\neg(\alpha\leftrightarrow\beta)$	$\alpha\leftrightarrow\neg\beta$
Connective translations	$\alpha\leftrightarrow\beta$	$(\alpha\rightarrow\beta)\wedge(\beta\rightarrow\alpha)$
	$\alpha\leftrightarrow\beta$	$(\alpha\wedge\beta)\vee(\neg\alpha\wedge\neg\beta)$
	$\alpha\rightarrow\beta$	$\neg(\alpha\wedge\neg\beta)$
	$\alpha\rightarrow\beta$	$\neg\alpha\vee\beta$
	$\alpha\vee\beta$	$\neg\alpha\rightarrow\beta$
Translations of negations	$\neg(\alpha\rightarrow\beta)$	$\alpha\wedge\neg\beta$
	$\neg(\alpha\wedge\beta)$	$\alpha\rightarrow\neg\beta$
	$\neg(\alpha\leftrightarrow\beta)$	$(\alpha\wedge\neg\beta)\vee(\beta\wedge\neg\alpha)$

EXERCISE 8.4.4

(a) Draw truth-tables to verify one of the de Morgan equivalences, one of the distribution principles, and association for $\leftrightarrow$.

(b) Verify the '$\neg$ through $\leftrightarrow$' equivalence in words, without writing out the table.

(c) An interesting feature of absorption, unlike the other equivalences listed in the table, is that the left will in general have fewer elementary letters than the right. Find two formulae that have no elementary letters in common but which are nevertheless tautologically equivalent. Comment on anything special you notice about them.

(d) Show from the definition that tautological equivalence is indeed an equivalence relation.

(e) Show that tautological equivalence is a *congruence relation* with respect to the propositional connectives, in the sense that whenever α -||- α' and β -||- β' then $\neg\alpha$, $\alpha\wedge\beta$, $\alpha\vee\beta$ are respectively equivalent to $\neg\alpha'$, $\alpha'\wedge\beta'$, $\alpha'\vee\beta'$.

(f) Consider the following *replacement property*. Let α be a formula, occurring as a subformula of γ. Suppose α -||- α'. Let γ' be formed by replacing this occurrence of α by α'. Show that γ -||- γ'. *Hint*: this is not as complicated as it sounds. It expresses the idea that we may interchange equivalent subformulae without loss of equivalence. Give two examples of the application of this property, and then sketch a proof of it, using the fact, already shown, that -||- is a congruence relation.

We saw analogues of many of these equivalences in the chapter on sets. For example, the de Morgan principles took the form of identities between sets: $-(A \cap B) = -A \cup -B$, $-(A \cup B) = -A \cap -B$, $A \cap B = -(-A \cup -B)$, and $A \cup B = -(-A \cap -B)$, where A, B, C are arbitrary sets and $-$ is complementation with respect to some local universe. This is not surprising: intersection, union and complementation of sets are defined using 'and', 'or' and 'not' respectively. There is thus a *systematic correspondence* between tautological equivalences and Boolean identities between sets. Similarly, there is a correspondence between tautological implications, such as those in Table 8.5, and inclusions between sets.

The de Morgan equivalences answer a question that we posed at the end of Section 8.2, about the connectives needed to be able to express all truth-functions. They may all be represented using just $\neg$, $\wedge$, since from those two we can get $\vee$ by the last of the four de Morgan equivalences, and we already know that with that trio we may obtain all the others. Likewise, the pair $\neg$, $\vee$ suffices to express all possible truth-functions, via the third de Morgan equivalence.

EXERCISE 8.4.5

(a) Use information from the list of equivalences to show that the pair $\neg, \rightarrow$ is enough to express all truth-functions.

(b) None of the three connectives $\neg$, $\wedge$, $\vee$ taken alone is sufficient to generate all truth-functions. However, there are just two among the 16 two-place truth-functions which can do the job alone. Go through the table for all these sixteen truth-functions that was constructed in an exercise for Section 8.3, use your intuition to pick likely candidates, and check that they do the job. *Hint*: First try to express $\neg$, and then try to express either $\wedge$ or $\vee$. If you succeed, you are done, for we know that negation and either one of conjunction, disjunction suffices. If you fail, try another candidate.

The concept of tautological equivalence may evidently be lifted to a relation between sets of formulae. If A, B are sets of formulae, we say that they are *tautologically equivalent* and write A -||- B iff A |- β for all $\beta \in B$ and also B |- α for all $\alpha \in A$. Equivalently: for every valuation v, $v(\alpha) = 1$ for all $\alpha \in A$ iff $v(\beta) = 1$ for all $\beta \in B$.

EXERCISE 8.4.6

(a) Check that tautological equivalence between sets of formulae is also an equivalence relation, and that singleton sets are equivalent iff their respective elements are so.

8.4.5 Tautologies and Contradictions

Now that we have the relations of tautological implication and equivalence under our belt, the properties of being a tautology, contradiction, or contingent are child's play. Let α be any formula.

- We say that α is a *tautology* iff $v(\alpha) = 1$ for every valuation v. In other words: iff α comes out with value 1 in every row of its truth-table.

- We say that α is a *contradiction* iff $v(\alpha) = 0$ for every valuation v. In other words: iff α comes out with value 0 in every row of its truth-table.

- We say that α is *contingent* iff it is neither a tautology nor a contradiction. In other words: iff $v(\alpha) = 1$ for some valuation v and also $v(\alpha) = 0$ for some valuation v.

Clearly, every formula is either a tautology, or a contradiction, or contingent, and only one of these. In other word, these three sets partition the set of all formulae into three cells.

EXERCISE 8.4.7 (WITH SOLUTION)

Classify the following formulae as tautologies, contradictions, or contingent: (i) $p \vee \neg p$, (ii) $\neg(p \vee \neg p)$, (iii) $p \vee \neg q$, (iv) $\neg(p \vee \neg q)$, (v) $(p \wedge (\neg p \vee q)) \rightarrow q$, (vi) $\neg(p \vee q) \leftrightarrow (\neg p \wedge \neg q)$, (vii) $p \wedge \neg p$ (viii) $p \rightarrow \neg p$, (ix) $p \leftrightarrow \neg p$, (x) $(r \wedge s) \vee \neg(r \wedge s)$, (xi) $(r \rightarrow s) \leftrightarrow \neg(r \rightarrow s)$.

Solution: Tautologies: (i), (v), (vi), (x), Contradictions: (ii), (vii), (ix). (xi). Contingent: (iii), (iv), (viii).

If you went through this exercise conscientiously, you will already have sensed many general lessons.

- A formula is a tautology iff its negation is a contradiction. Example: (i) and (ii).

- A formula is contingent iff its negation is contingent. Example: (iii) and (iv).

- We have $\alpha \vdash \beta$ iff the formula $\alpha \rightarrow \beta$ is a tautology. More generally, $\{\alpha_1, \ldots, \alpha_n\} \vdash \beta$ iff the formula $(\alpha_1 \wedge \ldots \wedge \alpha_n) \rightarrow \beta$ is a tautology. Example: the consequence $\{p, \neg p \vee q\} \vdash q$ (disjunctive syllogism) corresponds to the formula (v).

- We have $\alpha \dashv\vdash \beta$ iff the formula $\alpha \leftrightarrow \beta$ is a tautology. Example: the tautological equivalence $\neg(p \vee q) \dashv\vdash (\neg p \wedge \neg q)$ (de Morgan) corresponds to the formula (vi).

EXERCISE 8.4.8

Prove each of the bulleted points from the definitions.

Examples (x) and (xi) of the penultimate exercise are also instructive. The former tells us that $(r \wedge s) \vee \neg(r \wedge s)$ is a tautology. Without making a truth-table, we can see that it must be so since it is merely a substitution instance of the tautology $p \vee \neg p$. Likewise, $(r \rightarrow s) \leftrightarrow \neg(r \rightarrow s)$ is a contradiction, being a substitution instance of the contradiction $p \leftrightarrow \neg p$. Quite generally, we have the principle:

- Every substitution instance of a tautology or a contradiction is a tautology or contradiction, respectively.

On the other hand, not every substitution instance of a contingent formula is contingent. For example, we saw that $p \vee \neg q$ is contingent, but its substitution instance $p \vee \neg p$ (formed by substituting p for q) is a tautology, while another of its substitution instances $(p \wedge \neg p) \vee \neg(q \vee \neg q)$, formed by substituting $p \wedge \neg p$ for p and $q \vee \neg q$ for q, is a contradiction.

This notion of a substitution in propositional logic can be given a very precise mathematical content. A *substitution function* is a function $\sigma: L \rightarrow L$, where L is the set of all formulae, satisfying the following *homomorphism conditions*:

$$\sigma(\neg\alpha) = \neg\sigma(\alpha)$$

$$\sigma(\alpha \wedge \beta) = \sigma(\alpha) \wedge \sigma(\beta)$$

$$\sigma(\alpha \vee \beta) = \sigma(\alpha) \vee \sigma(\beta)$$

Note that in this definition, $=$ is not just tautological equivalence, it is full identity between formulae, i.e. the left and right sides stand for the very same formula. It is easy to show, by structural induction on the definition of a formula of propositional logic, that a substitution function is uniquely determined by its values for elementary letters.

EXERCISE 8.4.9 (WITH PARTIAL SOLUTION)

(a) Suppose $\sigma(p) = q \wedge \neg r$, $\sigma(q) = \neg q$, $\sigma(r) = p \rightarrow s$. Identify the formulae (i) $\sigma(\neg p)$, (ii) $\sigma(p \vee \neg q)$, (iii) $\sigma(r \vee (q \rightarrow r))$.

(b) Treating $\alpha \rightarrow \beta$ and $\alpha \leftrightarrow \beta$ as abbreviations for $\neg(\alpha \wedge \neg \beta)$ and $(\alpha \wedge \beta) \vee (\neg \alpha \wedge \neg \beta)$ respectively, show that substitution functions satisfy analogous homomorphism conditions for them too.

Solution to (a): (i) $\sigma(\neg p) = \neg \sigma(p) = \neg(q \wedge \neg r)$, (ii) $\sigma(p \vee \neg q) = \sigma(p) \vee \sigma(\neg q)$ $= \sigma(p) \vee \neg \sigma(q) = (q \wedge \neg r) \vee \neg \neg q$. For (iii), omitting the intermediate calculations, $\sigma(r \vee (q \rightarrow r)) = (p \rightarrow s) \vee (\neg q \rightarrow (p \rightarrow s))$.

Comment: Note that in e.g. (ii), $\sigma(p \vee \neg q)$ is obtained by *simultaneously* substituting $\sigma(p)$ for p and $\sigma(q)$ for q. If we were to replace *serially*, substituting $\sigma(p)$ for p to get $(q \wedge \neg r) \vee \neg q$ and then substituting $\sigma(q)$ for q, we would get the different result $(\neg q \wedge \neg r) \vee \neg \neg q$. Substitution as defined is always understood to be simultaneous.

We are now in a position to prove the claim that every substitution instance of a tautology is a tautology or, as we also say, the set of all tautologies is *closed* under substitution. Let α be any formula, and σ any substitution function. Suppose that $\sigma(\alpha)$ is not a tautology. Then there is a valuation v such that $v(\sigma(\alpha)) = 0$. Let v' be the valuation defined on the letters in α by putting $v'(p) = v(\sigma(p))$. Then it is easy to verify by structural induction that for every formula β, $v'(\beta) = v(\sigma(\beta))$. In particular, $v'(\alpha) = v(\sigma(\alpha)) = 0$, so that α is not a tautology.

Likewise, the set of all contradictions, and the relations of tautological equivalence and tautological implication are closed under substitution. That is, for any substitution function σ:

- Whenever α is a contradiction then $\sigma(\alpha)$ is too

- Whenever α -||- β then $\sigma(\alpha)$ -||- $\sigma(\beta)$

- Whenever α |- β then $\sigma(\alpha)$ |- $\sigma(\beta)$

- Whenever A |- β then $\sigma(A)$ |- $\sigma(\beta)$ for any substitution function σ.

Here A is any set of formulae, and $\sigma(A)$ is defined, as you would expect from the chapter on functions, as $\{\sigma(\alpha): \alpha \in A\}$.

EXERCISE 8.4.10 (WITH PARTIAL SOLUTION)

Complete the details of the verification that the set of all tautologies is closed under substitution.

8.5 Normal Forms, Least Letter-Sets, Greatest Modularity

The formulae of propositional logic can be of any length, depth or level of complexity. It is thus very important to be able to express them in the most transparent possible way. In this section we will look briefly at two kinds of normal form and two kinds of letter management. The two normal forms focus on the role of the connectives, allowing them to be applied in a predetermined order. Of the two kinds of letter management, one gets rid of all redundant letters, while the other modularizes their interaction as much as possible.

8.5.1 Disjunctive Normal Form

In logic, mathematics and computer science, a *normal form* (also often referred to as *canonical form*) for an expression is another one, equivalent to the first, but with a nice simple structure. A *normal form theorem* is one telling us that every expression (from some broad category) has a normal form (of some specified kind).

In propositional logic, the best-known such form is disjunctive normal form, abbreviated dnf. To explain it, we need the concepts of a literal and of a basic conjunction.

A *literal* is simply an elementary letter or its negation. A *basic conjunction* is any conjunction of (one or more) literals in which no letter occurs more than once. Thus we do not allow repetitions of a letter, as in $p \wedge q \wedge p$, nor repetitions of the negation of a letter, as in $\neg p \wedge q \wedge \neg p$, nor a letter occurring both negated and unnegated, as in $\neg p \wedge q \wedge p$. We write a basic conjunction as $\pm p_1 \wedge \ldots \wedge \pm p_n$, where the $\pm$ indicate the presence or absence of a negation sign.

A formula is said to be in *disjunctive normal form* (*dnf*) iff it is a disjunction of basic conjunctions. It is said to be a *full dnf* of a formula α iff every letter of α occurs in each of the (one or more) basic conjunctions.

EXERCISE 8.5.1 (WITH SOLUTION)

(a) Which of the following are in disjunctive normal form? When your answer is negative, explain briefly why. (i) $((p \wedge q) \vee \neg r) \wedge \neg s$, (ii) $(p \vee q) \vee (q \rightarrow r)$, (iii) $(p \wedge q) \vee (\neg p \wedge \neg q)$, (iv) $(p \wedge q) \vee (\neg p \wedge \neg q \wedge p)$, (v) $(p \wedge q) \vee (\neg p \wedge \neg q \wedge \neg p)$, (vi) $p \wedge q \wedge r$, (vii) p, (viii) $\neg p$, (ix) $p \vee q$, (x) $p \vee \neg p$, (xi) $p \wedge \neg p$.

(b) Which of the above are in full dnf?

Solution:

(a) No: there is a disjunction inside a conjunction, (ii) no: we have not eliminated $\rightarrow$, (iii) yes, (iv) no: $\neg p \wedge \neg q \wedge p$ contains two occurrences of p and so is not a basic conjunction, (v) no: $\neg p \wedge \neg q \wedge \neg p$ contains two occurrences of p, (vi) yes, (vii) yes, (viii) yes, (ix) yes, (x) yes, (xi) no: $p \wedge \neg p$ has two occurrences of p.

(b) (iii), (vi), (vii), (viii), (x).

How can we find a disjunctive normal form for an arbitrarily given formula α? There are two basic algorithms. One is semantic, proceeding via a truth-table for α. The other is syntactic, using successive transformations of α, justified by tautological equivalences from our table of basic equivalences.

The semantic construction is the simpler of the two. In fact we already made use of it in Section 8.3, when we showed that the trio $\neg, \wedge, \vee$ suffice for expressing all truth-functions. We begin by drawing up the truth-table for α, and checking whether it is a contradiction. Then:

- In the principal case that α is not a contradiction, the dnf of α is the disjunction of the basic conjunctions corresponding to the rows of the table in which α receives value 1.

- In the limiting case that α is a contradiction, i.e. when there are no such rows, the formula does not have a dnf.

It is clear from the construction that *every non-contradictory formula has a disjunctive normal form*. It is also clear that the dnf obtained is *unique* up to the ordering and bracketing of literals and basic conjuncts. Moreover, it is clearly *full*.

EXERCISE 8.5.2

Find the full disjunctive normal form (if it exists) for each of the following formulae, using the truth-table algorithm above: (i) $p \leftrightarrow q$, (ii) $p \rightarrow (q \vee r)$, (iii) $\neg[(p \wedge q) \rightarrow \neg(r \vee s)]$, (iv) $p \rightarrow (q \rightarrow p)$, (v) $(p \vee \neg q) \rightarrow (r \wedge \neg s)$, (vi) $(p \vee \neg p) \rightarrow (r \wedge \neg r)$.

For the *syntactic method*, we start with the formula α and proceed by a series of transformations that massage it into the desired shape. The basic idea is fairly simple, although the details are rather fussy. The following steps of the algorithm should be executed in the order given.

- Translate the connectives $\rightarrow$ and $\leftrightarrow$ (and any others in the language, such as exclusive disjunction) into $\neg, \wedge, \vee$ using translation equivalences such as those in Table 8.6.

- Use the de Morgan rules $\neg(\alpha\wedge\beta)$ -||- $\neg\alpha\vee\neg\beta$ and $\neg(\alpha\vee\beta)$ -||- $\neg\alpha\wedge\neg\beta$ iteratively, to move negation signs inwards until they act directly on elementary letters, eliminating double negations as you go by the rule $\neg\neg\alpha$ -||- α.

- Use the distribution rule $\alpha\wedge(\beta\vee\gamma)$ -||- $(\alpha\wedge\beta)\vee(\alpha\wedge\gamma)$ to move all conjunctions inside disjunctions.

- Use absorption and idempotence, with help from commutation and association, to eliminate repetitions of letters or of their negations.

- Delete all basic conjunctions containing a letter and its negation.

This will give us as output a formula in disjunctive normal form, except when α is a contradiction, in which case it turns out that the output is empty. However, the dnf obtained in this way is rarely full: there may be basic conjunctions with less than the full baggage of letters occurring in them.

EXERCISE 8.5.3

(a) Use the syntactic algorithm to transform the formula $(p\vee\neg q)\rightarrow(r\wedge\neg s)$ into disjunctive normal form. Show your transformations step by step, and compare the result with that obtained by the semantic method in the preceding exercise.

(b) How might you transform the dnf obtained above into a full one? *Hint*: make use of the tautological equivalence of expansion, α -||- $(\alpha\wedge\beta)\vee(\alpha\wedge\neg\beta)$.

8.5.2 Conjunctive Normal Form

Conjunctive normal form is like disjunctive normal form but 'upside-down': the roles of disjunction and conjunction are reversed. Technically, they are called *duals* of each other.

A *basic disjunction* is defined to be any disjunction of (one or more) literals in which no letter occurs more than once. A formula is said to be in *conjunctive normal form* (*cnf*) iff it is a conjunction of (one or more) basic disjunctions, and this is said to be a *full* conjunctive normal form of α iff every letter of α occurs in each of the basic disjunctions.

Cnfs may also be constructed semantically, syntactically, or by piggybacking on dnfs. For the semantic construction, we look at the rows of the truth-table that give α the value 0. For each such row we construct a basic disjunction: this will be the disjunction of those letters with the value 0 and the negations of the letters with value 1. We then conjoin these basic disjunctions. This will give an output in conjunctive normal form except when the initial formula α is a tautology. For the syntactic construction, we proceed just as we do for dnfs, except that we use the other distribution rule $\alpha \vee (\beta \wedge \gamma)$ -||- $(\alpha \vee \beta) \wedge (\alpha \vee \gamma)$ to move disjunctions inside conjunctions.

Alice Box: Conjunctive normal form

Alice: OK, I see *what* we are doing, but I don't see *why* we are doing it. Why do we construct the cnf from the table the way described? What is the underlying idea?

Hatter: We want to exclude the valuations that make α false. To do that, we take each row that gives α the value 0, and declare it not to hold. This is the same as negating the basic conjunction of that row, and by de Morganizing that negation we get the basic disjunction described. In effect, the basic disjunction says 'not this row'.

Alice: Why should we bother with cnfs when we already have dnfs? They are much less intuitive!

Hatter: They have been found useful in a discipline known as *logic programming*, for two reasons. First, a *conjunction* of basic disjunctions may be rewritten, without loss of equivalence, as a *set* of them, which helps us break problems down into smaller pieces. Second, in the special case of a basic disjunction with just one unnegated letter, say $(\neg p_1 \vee \ldots \neg p_n) \vee q$ $(n \geq 0)$, we may express it equivalently as a *positive implication* $(p_1 \wedge \ldots \wedge p_n) \to q$, understood as just the letter q in the limiting case that $n = 0$. If we have a set of these, we can programme a computer to derive conclusions simply by successive applications of modus ponens. But that takes us far beyond our present concerns.

We can also construct a cnf for a formula α by piggy-backing on a dnf for its negation $\neg \alpha$. Given α, construct a dnf of its negation $\neg \alpha$, by whatever method you like. This will be a disjunction $\beta_1 \vee \ldots \vee \beta_n$ of basic conjunctions β_i. Negate it, getting $\neg(\beta_1 \vee \ldots \vee \beta_n)$, which by double negation is evidently tautologically equivalent to α. We can then use de Morgan to push all negations inside, getting $\neg \beta_1 \wedge \ldots \wedge \neg \beta_n$. Each $\neg \beta_i$ is of the form $\neg(\pm p_1 \wedge \ldots \wedge \pm p_n)$, so we can apply de Morgan again, with double negation as needed, to express it as a disjunction of literals, as desired. If you compare this 'indirect' method with the semantic one for finding a cnf, you can see how it is doing much the same thing, without looking directly at a truth-table.

EXERCISE 8.5.4

(a) Take again the formula $(p \lor \neg q) \to (r \land \neg s)$, and find a cnf for it by all three methods: (i) semantic, (ii) successive syntactic transformations, (iii) the indirect method.

(b) Evidently, in most cases a formula that is in dnf will not be in cnf. But in some limiting cases it will be in both forms. When can that happen?

(c) Check that $\neg p_1 \lor \ldots \neg p_n \lor q$ -||- $(p_1 \land \ldots \land p_n) \to q$.

8.5.3 Eliminating Redundant Letters

A formula of propositional logic may contain redundant letters. We have already seen an example in the table of important equivalences: absorption tells us that $p \land (p \lor q)$ -||- p -||- $p \lor (p \land q)$; the letter q is thus redundant in each of the two outlying formulae. The expansion equivalences in the same table give another example. So does the equivalence $(p \to q) \land (\neg p \to q)$ -||- q, not in the table.

Suppose we expand our language a little to admit a zero-ary connective $\bot$ (called *bottom*), so that $\bot$ is a formula with no elementary letters. We stipulate that it receives the value 0 under every valuation. We also stipulate that $\sigma(\bot) = \bot$ for any substitution function σ.

Then contradictions like $p \land \neg p$ and tautologies like $p \lor \neg p$ will also have redundant letters, since $p \land \neg p$ -||- $\bot$ and $p \lor \neg p$ -||- $\neg \bot$. For our discussion of redundancy, we will assume that our language contains $\bot$ as well as the ordinary elementary letters, as it will streamline formulations.

It is clear that for every formula α containing elementary letters $p_1,..,p_n$ $(n \geq 0)$ there is a *minimal* set of letters in terms of which α may equivalently be expressed. That is, there is some minimal set F of letters such that α-||-α' for some formula α' all of whose letters are drawn from F. This is because α contains only finitely many letters to begin with, and so (by induction) as we eliminate redundant letters we must eventually come to a set (perhaps empty) from which no more can be eliminated.

But is this minimal set unique? In other words, is there a *least* such set – one that is included in every such set? In general, as noted in an exercise at the end of the chapter on relations, minimality does not in general imply leastness. However, in the present context, intuition suggests that surely there should be a least letter-set. And in this case intuition is right, although the proof is a little too advanced to give in this text.

In other words, we have the following theorem: For every formula α there is a *unique least* set E_α of letters having the property that α -| |- α' for some formula α' all of whose letters are drawn from E_α. This set E_α is known as *the least letter-set* for α, and any formula α' built from those letters and equivalent to α is known as a *least letter-set version* of α. The same considerations apply for arbitrary sets of formulae, even when they are infinite: each set A of formulae has a unique least letter-set.

It is possible to construct algorithms to find a least letter-set version of any formula, although they will in general be horribly exponential. For human work with small examples, we can use our experience with truth-functional formulae to inspire our guessing, backing this up by checking, as in the following exercise.

EXERCISE 8.5.5 (WITH SOLUTION)

(a) Find a least letter-set version for each of the following formulae: (i) $p \wedge \neg p$, (ii) $(p \wedge q \wedge r) \vee (p \wedge \neg q \wedge r)$, (iii) $(p \leftrightarrow q) \vee (p \leftrightarrow r) \vee (q \leftrightarrow r)$.

(b) Do the same for the following sets of formulae: (i) $\{p \vee q \vee r,\ p \vee \neg q \vee r\}$, (ii) $\{p \rightarrow q,\ p \rightarrow \neg q\}$.

(c) True or false? 'The least letter-set of a finite set of formulae is the same as the least letter-set of the conjunction of all of its elements'. Explain.

Solution:

(a) (i) $\bot$, (ii) $p \wedge r$, (iii) $\neg \bot$. (b) (i) $p \wedge r$, (ii) $\neg p$. (c) True: A finite set of formulae is tautologically equivalent to the conjunction of all of its elements, so the formulae that are equivalent to the former are the same as those equivalent to the latter.

8.5.4 Most Modular Representation

The next kind of simplification is rather more subtle. It leads us to a representation that does not change the letter-set, but makes their role as modular as possible. The definition makes essential use of the notion of a partition, and you are advised to review the basic theory of partitions in the chapter on relations before going further.

Consider the formula set $A = \{\neg p,\ r \rightarrow ((\neg p \wedge s) \vee q),\ q \rightarrow p\}$. This has three formulae as elements. Between them the formulae contain four elementary letters p, q, r, s. None of these letters is redundant – the least letter-set is still $\{p, q, r, s\}$. But the way in which the letters occur in formulae in A is unnecessarily 'mixed up': they can be separated out rather better from each other. In other words, we can make the presentation of A more 'modular', without reducing the set of letters involved.

Observe that A is tautologically equivalent to the set $A' = \{\neg p,\ r \rightarrow s,\ \neg q\}$. We have not eliminated any letters, but we have disentangled their role in the set. In effect, we have partitioned the letter set $\{p, q, r, s\}$ of A into three cells $\{p\}$, $\{r,s\}$,

$\{q\}$, with each formula in A' drawing all its letters from a single cell of the partition. Thus the formula $\neg p$ takes its sole letter from the cell $\{p\}$; $r \rightarrow s$ draws its letters from the cell $\{r,s\}$; and $\neg q$ takes its letter from the cell $\{q\}$. We say that the partition $\{\{p\}, \{r,s\}, \{q\}\}$ of the letter-set $\{p, q, r, s\}$ is a *splitting* of A.

In general terms, here is the definition. Let A be any set of formulae, with E the set of all its elementary letters. A *splitting* of A is a partition of E such that A is tautologically equivalent to some set A' of formulae, such that each formula in A' takes all of its letters from a single cell.

Now, as we saw in the chapter on relations, partitions of a set can be compared according to their fineness. Consider any two partitions of the same set. One partition is said to be at least as *fine* as another iff every cell of the former is a subset of some cell of the latter. This relation between partitions is a partial ordering (in the sense defined in the chapter on relations: reflexive, transitive, antisymmetric). As we also saw in the exercises at the end of that chapter, one partition is at least as fine as another iff the equivalence relation corresponding to the former is a sub-relation of that corresponding to the latter.

Since a splitting is a special kind of partition, it makes sense to compare splittings according to their fineness. Thus the three-cell splitting $\{\{p\}, \{r,s\}, \{q\}\}$ mentioned above is finer than the two-cell splitting $\{\{p,q\}, \{r,s\}\}$ corresponding to the formula set $A' = \{\neg p \wedge \neg q, r \rightarrow s\}$ equivalent to A; and this is in turn finer than the one-cell splitting $\{\{p, q, r, s\}\}$ corresponding to A itself.

It turns out that each set A of formulae has a *unique finest splitting* of its letter-set. In the case of our example, it is the three-cell partition given. The four-cell partition $\{\{p\}, \{r\}, \{s\}, \{q\}\}$ is finer – but it is not a splitting of A, since there is no equivalent set A'' of formulae, each of which draws its letters from a single cell of this partition (so that each formula in A'' contains only a single letter).

Any formula set A' equivalent to A, using the same letters as A, but with the letters in each formula of A' taken from a single cell of the finest splitting of A, is called a *finest splitting version* of A.

Strictly speaking, the definition above covers only the principal case that there is some elementary letter in some formula of A. For in the limiting case that there are no elementary letters, i.e. that $E = \varnothing$, the notion of a partition of E is not defined. However, in this case A must be tautologically equivalent either to $\perp$ or to $\neg\perp$ (as can be shown by an easy structural induction) and it is convenient to take that as the finest splitting version of A.

Thus a finest splitting version of a set of formulae disentangles as much as possible the roles that are played by the different elementary letters. It makes the presentation of the set as modular as possible: we have reached the finest way of separating the letters such that no formula contains letters from two distinct cells. The finest splitting versions of A may thus also be called *the most modular presentations* of A. They could also be called the *minimal mixing* versions of A.

Whereas the idea of the least letter-set of a set of formulae is quite old, perhaps dating back as far as the nineteenth century, that of the finest splitting is surprisingly new. It was first formulated and verified for the finite case by Rohit Parikh in 1999; a proof of uniqueness for the infinite case was given only in 2007!

As for finding least letter-set versions, any general algorithm is highly exponential. So for small examples to be analysed by hand, we again use experience with truth-functional formulae to inspire our guessing and follow it up by checking, in the following exercise.

EXERCISE 8.5.6 (WITH SOLUTION)

Find most modular versions for the following two sets of formulae: $A = \{p \wedge q \wedge r\}$, $B = \{p \rightarrow q, \ q \rightarrow r, \ r \rightarrow \neg p\}$. *Remember:* you are not eliminating redundant letters (indeed, in these instances no letters are redundant); you are splitting the existing letter-set.

Solution with comments: A most modular version of A would be $A' = \{p, q, r\}$. *Comment*: A is a singleton, and its unique element is a conjunction. We have simply broken the conjunction into its separate conjuncts, giving us the finest splitting $\{\{p\}, \{q\}, \{r\}\}$ with singleton cells.

A most modular version of B would be $B' = \{\neg p, \ q \rightarrow r\}$. *Comment*: The splitting given by B itself was the one-cell partition $\{\{p,q,r\}\}$, while B' gives us the finer two-cell partition $\{\{p\}, \{q, r\}\}$.

Finally, we note that there is nothing to stop us combining these two kinds of letter management. Given a set A of formulae we can first eliminate redundant letters, getting it into least letter-set form A', and then work on that to modularize the representation, obtaining a most modular version A'' of the least letter-set version. We can then go on, if we wish, to express the separate formulae $\alpha \in A''$ in dnf or in cnf.

8.6 Semantic Decomposition Trees

By now, you are probably sick of drawing up truth-tables, even small ones of four or eight rows, to test for the various kinds of tautologicality (tautological consequence and equivalence, tautologies and contradictions). Indeed, you may have tried some informal shortcuts to do the job. The final sections of this chapter show how shortcuts can be used systematically and in good conscience to great advantage. We sketch two well-known methods:

- A *semantic* one (i.e. formulated in terms of truth-values), that is *two-sided*, in the sense that it gives a way of determining whether or not a formula is e.g. a tautological consequence of others. As applications of the method always terminate in a finite time, this means that it supplies us with a *decision procedure* for tautologicality. It is called the method of *semantic decomposition trees* (also known as *semantic tableaux*), and will be explained in this section.

- A *syntactic* one (i.e. without direct reference to truth-values), that is *one-sided*, in the sense that it can eventually (and sometimes quickly) give a positive answer. But never, by itself, gives a negative one. Thus it does not provide a decision procedure, but has other advantages. This is the method of *natural deduction*, with its two components *enchainment* and *second-level* (alias *indirect*) inference. It will be sketched in the following section.

We begin with an example. We already know that the formula $\alpha = \neg(p \wedge q) \rightarrow (\neg p \vee \neg q)$ is a tautology, but let's check it without making a table. Suppose that $v(\alpha) = 0$ for some valuation v. Then $v(\neg(p \wedge q)) = 1$ while $v(\neg p \vee \neg q) = 0$. From the latter by the table for disjunction, $v(\neg p) = 0$ and $v(\neg q) = 0$, so by the table for negation, $v(p) = 1$ and $v(q) = 1$. On the other hand, since $v(\neg(p \wedge q)) = 1$ we have $v(p \wedge q) = 0$ so by the table for conjunction, either $v(p) = 0$ or $v(q) = 0$. In the first case we get a contradiction with $v(p) = 1$ and in the second case a contradiction with $v(q) = 1$. Thus the initial supposition that $v(\alpha) = 0$ is impossible, so $\neg(p \wedge q) \rightarrow (\neg p \vee \neg q)$ is a tautology.

This reasoning can be set out in the form of a labelled tree, as in Figure 8.1. Note the following features of its construction:

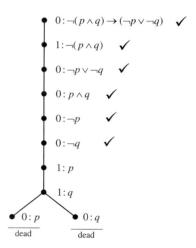

Figure 8.1   *Semantic decomposition tree for $\neg(p \wedge q) \rightarrow (\neg p \vee \neg q)$*

- The *root* is labelled with the formula that we are testing, together with a truth value. In our example, we are testing whether α is a tautology, i.e. whether it is impossible for it to receive value 0, so the label is 0: α. If we had been testing whether α is a contradiction, the label would have been 1: α.

- At each step we *decompose* the current formula, passing from information about its truth-value to resulting information about its *immediate subformulae*. Never in the opposite direction.

- When we get *disjunctive* information about the immediate subformulae of φ, we *divide* our branch into two sub-branches, with one of the alternatives on one branch and the other alternative on the other.

- When we get *definite* information about the immediate subformulae of φ, we put it on *every branch* below the node for φ. One way of ensuring this is by putting it before any further dividing takes place.

Note also the way in which we read our answer from the tree thus constructed. We make sure that we have decomposed each node in the tree whenever it is possible, which is whenever it is not an elementary letter. In our example, the ticks next to nodes keep a record that we have done the decomposition. Then we read the completed tree as follows:

- If *every* branch contains an *explicit contradiction* (two nodes labelled by the same formula with opposite signs, 1: φ and 0: φ) then the label of the root is impossible. This is what happens in our example: there are two branches, one containing both $v(p) = 1$ and $v(p) = 0$, the other containing both $v(q) = 1$ and $v(q) = 0$. We label these branches *dead* and conclude that $v(\alpha) = 0$ is impossible, i.e. that α is a tautology.

- On the other hand, if *some* branch is without explicit contradiction, then the label of the root is possible. We label these branches *ok*, and read off any one of them a valuation that gives the root formula the value indicated in the label.

We give a second example that illustrates the latter situation. Consider the formula $\alpha = ((p \land q) \to r) \to (p \to r)$, and test to determine whether or not it is a tautology. We get the labelled tree of Figure 8.2.

This decomposition tree contains three branches. The leftmost and rightmost ones each contain an explicit contradiction (1: p and 0: p in the left one, 0: r and 1: r in the right one) and so we label them *dead*. The middle branch does not contain any explicit contradiction. We check carefully that we have completed all decompositions in this branch – that we have ticked every formula in it other than elementary letters. We collect from the branch the assignment $v(p) = 1$, $v(r) =$

$v(q) = 0$. This valuation generated by this assignment gives every formula in the branch its labelled value, and in particular gives the root formula α the labelled value 0, so that it is not a tautology.

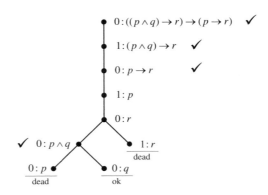

Figure 8.2 *Semantic decomposition tree for $((p \wedge q) \to r) \to ((p \to r)$*

Evidently, this method is algorithmic and can be programmed. It is usually quicker to calculate (whether by human or by machine) than a full table, although it must be admitted that in worst-case examples, it turns out to be just as horribly exponential as a truth-table. However, even in such cases, it is rather more fun to carry out (for humans, at least).

The construction always terminates: as we go down branches we deal with shorter and shorter formulae until we reach elementary letters and can decompose no further. This is intuitively clear: a rigorous proof would be by structural induction on formulae.

There are many ways in the method can be streamlined to maximize its efficiency. Some gains may be obtained by controlling the order in which decompositions are carried out. In our second example, after decomposing the root to introduce the second and third nodes, we had a choice of which of those to decompose first. We opted to decompose the third node before the second, but we could perfectly well have done the reverse. Our choice was motivated by the fact that the third node gives definite information, following a policy of postponing branching for as long as possible. This is a control heuristic that works well, at least for humans in small finite examples. Whichever order we follow, we get the same final verdict.

EXERCISE 8.6.1

Reconstruct the semantic decomposition tree for $((p \wedge q) \to r) \to (p \to r)$, decomposing the second node before the third one.

More important as a way of reducing unnecessary work is the fact that we can sometimes stop constructing the tree before it is finished. If a branch contains an explicit contradiction before having been fully decomposed, we can already declare it dead and pass on to other branches. Also, if we have a branch that is complete in the sense that all nodes on that branch have been decomposed, and it is free of explicit contradictions, then we can declare that this branch is *ok* and the label of the root (and of everything on the branch) is possible, without bothering to complete any other branches.

In addition to these economies, we can also develop *heuristics* to guide the choice which branches to construct first and to what extent, balancing depth-first and breadth-first strategies. But all of this goes far beyond the limits of this introduction, and we leave it aside. Tables 8.8 and 8.9 recapitulate the definite and disjunctive decompositions that one may make with the various truth-functional connectives. Their justification is immediate from the truth-tables. With it in hand, the following exercise should be straightforward.

Table 8.8 Definite decomposition rules.

1: ¬α	0: ¬α	1: α∧β	0: α∨β	0: α→β
0: α	1: α	1: α	0: α	1: α
		1: β	0: β	0: β

Table 8.9 Disjunctive decomposition rules.

1: α∨β		0: α∧β		1: α→β		1: α↔β		0: α↔β	
1: α	1: β	0: α	0: β	0: α	1: β	1: α	0: α	1: α	0: α
						1: β	0: β	0: β	1: β

We thus have two rules for decomposing each connective – one for sign 1 and the other for sign 0. Note carefully the rule for decomposing material implication. It is evident from the corresponding truth-table, but is often misremembered. Note that the rule for negation is definite no matter what the sign (1 or 0), while the one for material equivalence is disjunctive irrespective of sign. In the latter case, however, we need to make two entries on each side of the fork.

EXERCISE 8.6.2

(a) What would the decomposition rules for exclusive disjunction look like? Would they be definite or disjunctive? Can you see any connection with the rules for any of the other connectives?

(b) Determine the status of the following formulae by the method of decomposition trees. First test to see whether it is a tautology. If it is not a tautology, test to see whether it is a contradiction. Your answer will specify one of the three possibilities: tautology, contradiction, contingent.

   (i) $(p{\rightarrow}(q{\rightarrow}r)){\rightarrow}((p{\rightarrow}q){\rightarrow}(p{\rightarrow}r))$

   (ii) $(p{\rightarrow}q){\vee}(q{\rightarrow}p)$

   (iii) $(p{\leftrightarrow}q){\vee}(q{\leftrightarrow}p)$

   (iv) $(p{\leftrightarrow}q){\vee}(p{\leftrightarrow}r){\vee}(q{\leftrightarrow}r)$

   (v) $(p{\rightarrow}q){\wedge}(q{\rightarrow}r){\wedge}(q{\rightarrow}{\neg}p)$

   (vi) $((p{\vee}q){\wedge}(p{\vee}r)){\rightarrow}(p{\wedge}(q{\vee}r))$

Clearly, the method may also be applied to determine whether two formulae are tautologically equivalent, whether on formula tautologically implies another, and more generally whether a finite set of formulae tautologically implies a formula.

- To test whether α -||- β, test whether the formula $\alpha{\leftrightarrow}\beta$ is a tautology.

- To test whether α |- β, test whether the formula $\alpha{\rightarrow}\beta$ is a tautology.

- To test whether $\{\alpha_1,\ldots,\alpha_n\}$ |- β, test whether the formula $(\alpha_1{\ldots}\alpha_n){\rightarrow}\beta$ is a tautology.

EXERCISE 8.6.3

(a) Use the method of decomposition trees to determine whether (i) $(p{\vee}q){\rightarrow}r$ -||- $(p{\rightarrow}r){\wedge}(q{\rightarrow}r)$, (ii) $(p{\wedge}q){\rightarrow}r$ -||- $(p{\rightarrow}r){\vee}(q{\rightarrow}r)$, (iii) $p{\wedge}{\neg}p$ -||- $q{\leftrightarrow}{\neg}q$.

(b) Use the method of decomposition trees to determine whether $\{{\neg}q{\vee}p,$ ${\neg}r{\rightarrow}{\neg}p,\ s,\ s{\rightarrow}{\neg}r,\ t{\rightarrow}p\}$ |- ${\neg}t{\wedge}{\neg}q$.

8.7 Natural Deduction

The second shortcut method puts onto a formal footing the procedures that mathematicians use in informal deductive inference. It is particularly well adapted to showing that a formula β is a tautological consequence of a set $\{\alpha_1,\ldots,\alpha_n\}$ of formulae, but may be used to check the other tautologicality relations as well. Its great attraction is that it is quite natural to anyone with some experience in mathematical reasoning (hence its name), while its drawback is that that in the negative case, when $\{\alpha_1,\ldots,\alpha_n\}$ |- β, it does not provide a definite answer. An attempt to establish a positive result may fail because it does not hold, or because the attempt was not sufficiently persistent or clever. This is in contrast to the method of semantic decomposition trees which, like entire truth-tables, always supplies an answer in a finite time.

Natural deduction has two parts: *enchainment* and *second-level* (or *indirect*) inference. We will sketch their essential ideas and look at some examples.

8.7.1 Enchainment

Consider an inference like the following, where we make every step explicit. We are given the premises $(p\vee\neg q)\rightarrow(s\rightarrow r)$, $\neg s\rightarrow q$, and $\neg q$. We want to prove r. We can do this by putting together simple tautological entailments to go step by step from the premises to the conclusion.

Specifically, from the third premise we have by addition $p\vee\neg q$, and from this together with the first premise by modus ponens $s\rightarrow r$. But from the third premise again, combined this time with the second one, we have by modus tollens $\neg\neg s$, which evidently by double negation elimination gives us s. Combining this with $s\rightarrow r$ gives us by modus ponens r, as desired.

Mathematically, the clearest way of setting this out is as a labelled derivation tree, as follows.

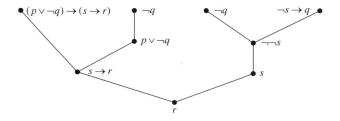

Figure 8.3   *A labelled derivation tree.*

Such a tree is conventionally written with leaves at the top, and is usually constructed from the leaves to the root. The leaves are labelled by the premises of the inference. The parent of a node (or pair of nodes) is labelled by the proposition immediately inferred from it/them. If desired, we may also label each node with the usual name of the inference rule used to get it or, in the case of the leaves, give it the label 'premise'. Our example gives us a labelled 2-tree, but not quite a binary tree since the single-child links are not labelled 'left' or 'right'. Note that the labelling function in the example is not quite injective: there are two distinct leaves labelled by the third premise.

Given the convention of linearity that is inherent in of our writing system (not only for English or just western languages, but all that have developed in different civilizations), an inference such as the above would not normally be written out as a tree, but as a finite *sequence*, perhaps with annotations, as in the table below. This is why we call it *enchainment*. The sequential presentation has the incidental advantage, over the tree presentation, of eliminating any need to repeat premises or other formulae.

Table 8.10 A derivation as a sequence with annotations.

Number	Formula	Obtained from	By rule
1	$(p\vee\neg q)\rightarrow(s\rightarrow r)$		premise
2	$\neg s\rightarrow q$		premise
3	$\neg q$		premise
4	$p\vee\neg q$	3	addition
5	$s\rightarrow r$	4, 1	modus ponens
6	$\neg\neg s$	3, 2	modus tollens
7	s	6	double negation
8	r	7, 5	modus ponens

EXERCISE 8.7.1

(a) In Figure 8.3 label the interior nodes with the rules used.

(b) What happens to the tree in Figure 8.3 if we merge the two nodes labelled $\neg q$?

(c) In the last exercise we used the method of decomposition trees to show that $\{\neg q\vee p,\ \neg r\rightarrow\neg p,\ s,\ s\rightarrow\neg r,\ t\rightarrow p\}\ |\text{-}\ \neg t\wedge\neg q$. Do the same by natural deduction, setting out the result first as a labeled derivation tree, and then as an annotated derivation sequence.

In our examples we have mainly used rules from the table of simple tautological implications, and only one (double negation) from the table of tautological equivalences. Evidently, we can use any of them too, and indeed in a stronger way. This is because of the *replacement property* for tautological equivalence, established in an exercise earlier in the chapter. We recall what it says: when α is a subformula of γ, and α -||- α', then γ -||- γ', where γ' is formed by replacing one occurrence of α by α'.

With this property we may pass from, say, $(p\wedge q)\rightarrow r$ directly to $\neg(\neg p\vee\neg q)\rightarrow r$ since the latter is obtained by replacing a subformula (the antecedent) of the former by one to which it is equivalent (by de Morgan). Compare this with passage from the same $(p \wedge q) \rightarrow r$ to $(p \wedge (q \vee s)) \rightarrow r$. This is invalid despite the fact that the latter formula is obtained from the former by replacing a subformula (the antecedent) by one which it tautologically implies.

8.7.2 Second-Level (Alias Indirect) Inference

The other component of natural deduction is *second-level inference*, also often known as *indirect inference*. Its characteristic feature is that we are not simply inferring the truth of one proposition from another; we are inferring the *validity of an entire inference* from that of one or more others. The general form of a second-level inference, in the context of propositional logic, is thus passage from k tautological implications, called *subordinate* inferences:

$$\alpha_{11}, \ldots \vdash \alpha_1$$
$$\cdots$$
$$\alpha_{k1}, \ldots \vdash \alpha_k$$

to a tautological implication, called the *principal* inference:

$$\beta_1, \ldots \vdash \beta.$$

We have already made use of three such principles in the intuitive reasoning of earlier chapters, describing them informally in logic boxes. They are: *conditional proof, disjunctive proof* and *proof by contradiction* (alias *reductio ad absurdum*). Systems of natural deduction formalize and codify their use. The three principles may are expressed in the following table, where the subordinate inferences are in the upper row and the corresponding principal ones are in the lower row. In each rule, $n \geq 0$.

Table 8.11 Three kinds of second-level inference.

Conditional Proof	Disjunctive Proof	Proof by Contradiction
$\alpha_1, \ldots, \alpha_n, \beta \vdash \gamma$	$\alpha_1, \ldots, \alpha_n, \beta_1 \vdash \gamma$	$\alpha_1, \ldots, \alpha_n, \neg\beta \vdash \gamma \wedge \neg\gamma$
	$\alpha_1, \ldots, \alpha_n, \beta_2 \vdash \gamma$	
$\alpha_1, \ldots, \alpha_n \vdash \beta \rightarrow \gamma$	$\alpha_1, \ldots, \alpha_n, \beta_1 \vee \beta_2 \vdash \gamma$	$\alpha_1, \ldots, \alpha_n \vdash \beta$

The reason why these three forms of inference are useful is they *give us more to grab hold of*:

- In conditional proof, we want to prove $\beta \rightarrow \gamma$ from n premises $\alpha_1, \ldots, \alpha_n$. The rule tells us that it suffices to prove the conclusion γ from $n+1$ premises $\alpha_1, \ldots, \alpha_n$, β – one more premise to play with! Incidentally, γ is also shorter than $\beta \rightarrow \gamma$, which also helps a bit.

- In disjunctive proof, we want to prove γ from $n+1$ premises $\alpha_1, \ldots, \alpha_n$, $\beta_1 \lor \beta_2$. The rule tells us that it suffices to prove the same conclusion γ from each of two premise-sets $\alpha_1, \ldots, \alpha_n$, β_1 and $\alpha_1, \ldots, \alpha_n$, β_2. The premises β_1 and β_2 of the two subordinate inferences are each in general *stronger* than the original disjunction $\beta_1 \lor \beta_2$, and so can be used to greater advantage.

- In proof by contradiction, we want to prove β from n premises $\alpha_1, \ldots, \alpha_n$. The rule tells us that it suffices to prove an explicit contradiction (any one will do) from the $n+1$ premises $\alpha_1, \ldots, \alpha_n, \neg\beta$. As with conditional proof, this gives us one more premise to deploy.

These are the three basic and most important forms of second-level inference. There are others that stem from them. For example, proof by contradiction may also be expressed as: from $\alpha_1, \ldots, \alpha_n, \beta \vdash \gamma \land \neg\gamma$ to $\alpha_1, \ldots, \alpha_n \vdash \neg\beta$. Conditional proof often takes the contraposed form: from $\alpha_1, \ldots, \alpha_n, \neg\gamma \vdash \neg\beta$ to $\alpha_1, \ldots, \alpha_n \vdash (\beta \rightarrow \gamma)$. But these are details.

Terminology: The additional premises of conditional proof, disjunctive proof and proof by contradiction are usually called *suppositions* (or *temporary assumptions*). A supposition is thus a premise of a subordinate inference that is not already one of the premises of the principal inference. The term 'indirect proof' is sometimes reserved for proof by contradiction, but we will use it broadly for all kinds of second-level inference.

Alice Box: Variant forms of indirect inference

Alice: Aren't the two forms of proof by contradiction exactly the same?

Hatter: Very nearly, but not quite; the position of the negation changes. Of course, as far as classical logic is concerned, they are trivially equivalent. Whatever inference we carry out with one, we can carry out using the other plus the tautological equivalence of double negation.

Alice: You say 'for classical logic'. Why?

Hatter: There are non-classical logics that abandon the principle of bivalence, and in which double negation fails. For them the two forms of proof by contradiction are not the same. In the case of so-called intuitionistic logic, for example, where double negation introduction holds but its elimination fails, the variant form of proof by contradiction holds but not our basic one. But all that is beyond our concerns...

Alice: It is a pity that disjunctive proof does not follow the same pattern as the other two, using an *additional* premise in each of the two subordinate proofs, rather than a *substitute* one.

(Continued)

> *Alice Box:* (Continued)
>
> *Hatter*: In fact, the rule can equivalently be formulated in that way: from $\alpha_1, \ldots, \alpha_n, \beta_1 \vee \beta_2, \beta_1 \mid\!\!- \gamma$ and $\alpha_1, \ldots, \alpha_n, \beta_1 \vee \beta_2, \beta_2 \mid\!\!- \gamma$ to $\alpha_1, \ldots, \alpha_n, \beta_1 \vee \beta_2 \mid\!\!- \gamma$. Here the premises of the two subordinate inferences are just those of the principal inference, *plus* a β_i. For uniformity, we could well regard this as the official form of disjunctive proof, and the earlier version as a streamlining of it. Without going into details, it should be clear that since each $\beta_i \mid\!\!- \beta_1 \vee \beta_2$, the two forms of disjunctive proof are equivalent.

Natural deduction is quite a Pandora's box, and some classes will want to call a halt at this point. The remaining few pages of this chapter are for those who would like to see just a little more about how it works.

Clearly, in any interesting indirect inference, we will have enchaining within the subordinate inferences. Moreover, once a second-level step is completed, we may wish to enchain its conclusion with other material. So a full account of natural deduction must look at how the two components work together, with conventions to permit complex inferences to be set out in as transparent and economical a manner as possible. Unfortunately, those two desiderata tend to conflict with each other. There are several strategic options available.

- From a mathematical point of view, working with trees, the most elegant approach is to give a recursive definition of a derivation tree under which leaves may be labelled not only by propositional formulae, *but also by already-constructed derivation trees*. For example, a leaf may be labelled by a derivation tree for a tautological consequence $\alpha, \neg\beta \mid\!\!- \gamma \wedge \neg\gamma$, and if another leaf is labelled by α, the two can give, as parent, a node labelled β.

- In practice, such recursive trees are usually merged into a *single tree*, in which additional labels are used to signify the 'cancellation' (also known as 'discharge') of the suppositions of subordinate inferences. This is the method used in most theoretical investigations stemming from a classic study by Dag Prawitz, and it works well for systems in which subordinate inferences have all the premises of the principal one, plus one.

- Introductory textbooks usually *linearize* everything, so that a complex inference involving both enchaining and second-level steps becomes a long enchainment, with lots of additional labelling to record the introduction and discharge of suppositions and keep track of dependencies. The conventions for displaying and annotating such a linearization are quite intricate, and every textbook of logic does it in a slightly different way, trying to balance the virtues of transparency, brevity, and naturalness.

We give three simple examples of how this may be done, with a fully linearized presentation and a very explicit system of annotation.

Example 1. We want to show that $\{(p\wedge q)\rightarrow r,\ (t\rightarrow\neg s)\wedge(r\rightarrow(t\vee u))\}$ |- $(p\wedge s)\rightarrow(q\rightarrow u)$. This contains 6 elementary letters, so a truth-table for it would have $2^6 = 64$ rows – too much to write easily by hand. A semantic decomposition tree would also be rather unwieldy. It is much more elegant to do it by natural deduction, applying the indirect rule of conditional proof twice. In the table below, the key entries, as far as indirect inference is concerned, are highlighted by boldface type. The stuff in between is just enchainment, which should by now be fairly routine.

Table 8.12 A natural deduction using conditional proof.

N°	Formula	From	Rule	Depends on	Current Goal
1	$(p\wedge q)\rightarrow r$		premise	1	
2	$(t\rightarrow\neg s)\wedge(r\rightarrow(t\vee u))$		premise	2	$(p\wedge s)\rightarrow(q\rightarrow u)$
3	**$p\wedge s$**		**supposition**	**3**	**$q\rightarrow u$**
4	**q**		**supposition**	**4**	**u**
5	p	3	simplification	3	ditto
6	$p\wedge q$	5,4	conjunction	3, 4	ditto
7	r	6, 1	modus ponens	1, 3, 4	ditto
8	$r\rightarrow(t\vee u)$	2	simplification	2	ditto
9	$t\vee u$	8, 7	modus ponens	1, 2, 3, 4	ditto
10	$t\rightarrow\neg s$	2	simplification	2	ditto
11	s	3	simplification	3	ditto
12	$\neg\neg s$	11	double negation	3	ditto
13	$\neg t$	12, 10	modus tollens	2, 3	ditto
14	u	13, 9	disj. syllogism	1, 2, 3, 4	**$q\rightarrow u$**
15	**$q\rightarrow u$**	**1,2,3,4** |- **14**	**conditional proof**	**1, 2, 3** (discharge 4)	**$(p\wedge s)\rightarrow(q\rightarrow u)$**
16	**$(p\wedge s)\rightarrow(q\rightarrow u)$**	**1,2,3,** |- **15**	**conditional proof**	**1, 2** (discharge 3)	□

Commentary.

- The first column simply numbers the steps for easy cross-reference. The second column gives us the *current formula* at each step in the proof, beginning with the premises (rows 1, 2) and ending with the desired conclusion (row 16). The remaining columns could be referred to disdainfully as mere 'book-keeping', but their entries are needed to ensure that there are no errors. Go through them carefully, with the following comments in mind.

- Column 3 tells us from what the current formula was *immediately inferred*. In particular:

  ○ Premises and suppositions are *not inferred* from anything. Hence the empty cells in rows 1–4.

○ When we apply an indirect inference, what we get is not inferred from a proposition, but from the *prior construction of a subordinate inference.* Hence the special entries in rows 15, 16.

• We need to keep track of the premises and suppositions on which the current formula *ultimately depends.* The ultimate dependencies are recorded in column 5. In particular:

○ With the application of enchainment rules, ultimate dependencies *grow cumulatively.*

○ But whenever a rule of indirect proof is applied, its supposition is *discharged,* i.e. the current formula no longer depends on it (rows 15, 16), thus *diminishing* the ultimate dependencies.

○ Suppositions are made and discharged on a last-in first-out basis. In computing terminology, the suppositions form a *stack.*

• It is vital to keep track of the *current goal.* This is constantly changing – in the example it changes four times! If we don't keep track of it, the proof wanders aimlessly. That is the purpose of column 6. In particular:

○ Under enchainment steps, the current goal remains unchanged until we reach it, so we write 'ditto' each time (rows 5–13). Of course, one could reduce repetition by simply leaving the ditto cells blank.

○ Making a supposition for conditional proof (rows 3, 4) or proof by contradiction changes the current goal. But making a supposition for a disjunctive proof does not.

○ When a supposition is discharged, the current goal reverts to what it was before that supposition was made. Hence the symmetry in column 6.

○ The last cell in column 6 contains an 'end of proof' sign, because we have reached our initial goal and so have no current goal left to pursue!

Example 2. We want to show that $\neg p \vee q$, $(\neg p \vee r) \rightarrow s$, $\neg s \rightarrow \neg q$ |- s using disjunctive proof. We do it in the next table.

Table 8.13 A natural deduction using iterated disjunctive proof.

N°	Formula	From	Rule	Depends on	Current Goal
1	$\neg p \vee q$		premise	1	
2	$(\neg p \vee r) \rightarrow s$		premise	2	
3	$\neg s \rightarrow \neg q$		premise	3	s
4	$\neg\boldsymbol{p}$		**supposition**	4	ditto
5	$\neg p \vee r$	4	addition	4	ditto

Table 8.13 (continued)

N°	Formula	From	Rule	Depends on	Current Goal
6	s	2, 5	modus ponens	2, 4	ditto
7	**q**		**supposition**	**7**	ditto
8	$\neg\neg q$	7,	double negation	7	ditto
9	$\neg\neg s$	3, 8	modus tollens	3, 7	ditto
10	s	9	double negation	3, 7	ditto
11	**s**	**2,4 |-s**	**disjunctive proof**	**1, 2, 3**	□
		3,7 |-s		**(discharge 4, 7)**	

Commentary. In row 4 we make the supposition $\neg p$, and use it with the help of 2 to get s in row 6, thus establishing the subordinate inference 2,4 |- s. Likewise we make the supposition q in row 7, and use it with the help of 3 to get s in row 10, thus establishing the subordinate inference 3,7 |- s. We may now apply the rule of disjunctive proof to get the same conclusion s without either of these two suppositions, but depending instead on the disjunctive premise 1 as well as on 2, 3.

Evidently in such derivations, applications of equivalences such as double negation, or commutation and association for $\wedge,\vee$, are rather tedious, and may well be effected jointly with neighbouring steps.

Example 3. We establish the consequence $(r\vee s)\rightarrow\neg p$ |- $\neg(p\wedge s)$ using proof by contradiction in the next table.

Table 8.14 A natural deduction using proof by contradiction.

N°	Formula	From	Rule	Depends on	Current Goal
1	$(r\vee s)\rightarrow\neg p$		premise	1	$\neg(p\wedge s)$
2	**$p\wedge s$**		**supposition**		**contradiction**
3	p	**2**	simplification	2	ditto
4	$\neg(r\vee s)$	1, 3	modus tollens, dn	1, 2	ditto
5	$\neg r\wedge\neg s$	4	de morgan	1, 2	ditto
6	$\neg s$	5	simplification	1, 2	ditto
7	s	2	simplification	2	ditto
8	$s\wedge\neg s$	7, 6	conjunction	1, 2	ditto
9	**$\neg(p\wedge s)$**	**1, 2 |- 8**	**proof by contradiction**	**1**	□
				(discharge 2)	

Commentary. Strictly speaking, for brevity we are using proof by contradiction in its variant form mentioned earlier. Moreover, it is rather pedantic to assemble $\neg s$ and s into a conjunction in row 8, and one could save the step by using the rule in a further equivalent guise: from $\alpha_1,\ldots,\alpha_n, \beta$ |- γ and $\alpha_1,\ldots,\alpha_n, \beta$ |- $\neg\gamma$ to $\alpha_1,\ldots,\alpha_n$ |- $\neg\beta$. The more one carries out natural deductions, the more such corners one likes to cut.

EXERCISE 8.7.2

(a) Solve the example in Table 8.1.2 using proof by contradiction.

(b) Do the same for the example in Table 8.1.3.

> *Comment*: This exercise illustrates the way in which proof by contradiction is a *universal option*, in the sense that it may always be applied, at any stage in any proof (indeed, even iterated within another such application). Some professionals love employing it, others prefer to avoid it.
>
> In contrast, conditional proof can only be applied when the desired conclusion is (or can be transformed equivalently into) conditional shape, and disjunctive proof is available only when one of the premises (or some formula obtained in the course of the derivation) presents itself as (or can be transformed into) a disjunction. Such transformations are always *possible*, but they can be rather artificial and not very helpful.

Although we have not spelled it out in detail, it should be apparent that the entire procedure of natural deduction makes essential (though implicit) use of the Tarski properties of reflexivity, cumulative transitivity and monotony of the relation of tautological consequence, which we outlined at the beginning of the chapter. Cumulative transitivity is what allows us to 'keep going' without loss of power in a derivation, and monotony permits us to take for granted that an inference from one or two of our premises is acceptable as an inference from all of them taken together.

FURTHER EXERCISES

8.1. *Structural properties of consequence*

(a) Show that the Tarski condition of monotony, for logical consequence as an operation, follows from the condition of monotony for the corresponding relation.

(b) Show that the Tarski condition of idempotence, for logical consequence as an operation, may be obtained from joint use of the conditions of reflexivity and cumulative transitivity for the corresponding relation.

8.2. *Truth-functional connectives*

(a) Construct the full disjunctive normal form for each of the two-place truth-functions f_2, f_{11} and f_{15} from the table in Section 8.3. Can any of them be expressed more simply?

(b) Show that the connectives $\wedge, \vee, \rightarrow, \leftrightarrow$ do not suffice to express all truth-functions. *Hint*: Think about the top row of the truth-table. If you write the proof out in any detail, at a certain stage you will need to carry out a structural induction on the set of all formulae that can be built with the mentioned connectives.

(c) Show that the pair $\neg$, $\leftrightarrow$ does not suffice to express all truth-functions. *Hint*: To do this, show that they do not suffice to express $\wedge$. To do that, first show that if they do suffice to express $\wedge$, then they can do so by a formula with just two elementary letters. Then show by structural induction that any formula with just two elementary letters and only $\neg$, $\leftrightarrow$ as connectives, is true in either 0, 2, or 4 of the rows of its truth-table.

8.3. *Tautologicality*

(a) Explain why the following two properties are equivalent: (i) α is a contradiction, (ii) $\alpha \vdash \beta$ for every formula β.

(b) Explain why the following three properties are equivalent: (i) α is a tautology, (ii) $\beta \vdash \alpha$ for every formula β, (iii) $\varnothing \vdash \alpha$.

(c) A set A of formulae is said to be *inconsistent* iff there is no valuation v such that $v(\alpha) = 1$ for all $\alpha \in A$. How does this concept, applied to finite sets of formulae, relate to that of a contradiction?

(d) Show, as claimed in the text, that the set of all contradictions and the relations of tautological equivalence and tautological implication, are closed under substitution. *Hint*: You may do this in either of two ways – either directly from their definitions, or from their connections with the property of being a tautology and the fact (shown in the text) that the set of all tautologies is closed under substitution.

(e) Show by structural induction that every formula constructed using the zero-ary connective $\perp$ and the usual $\neg, \wedge, \vee$ but without any elementary letters, is either a tautology or a contradiction.

8.4. *Normal forms*

(a) Find a dnf for each of the following formulae: (i) $r \rightarrow \neg(q \vee p)$, (ii) $p \wedge \neg(r \vee \neg(q \rightarrow \neg p))$, using the semantic method for one and the method of successive syntactic transformations for the other.

(b) Find a cnf for the formula $p \wedge \neg(r \vee \neg(q \rightarrow \neg p))$ by each of the three methods: (i) semantic, (ii) successive syntactic transformations, (iii) indirect method (via the dnf of its negation).

(c) For each of the following sets of formulae, first find a least letter-set version, and then find a finest splitting version of that: $C = \{p \leftrightarrow q, q \leftrightarrow r, r \leftrightarrow \neg p\}$, $D = \{(p \wedge q \wedge r) \vee s\}$, $E = \{(p \wedge q \wedge r) \vee (s \wedge q), \neg p\}$.

(d) True or false? For each, give proof or counter-example. (i) The least letter-set of a formula is empty iff the formula is either a tautology or a contradiction. (ii) If a letter is redundant in each of α, β then it is redundant in $\alpha \wedge \beta$. (iii) The least letter-set of a disjunction is the union of the least letter-sets of its disjuncts.

8.5. *Semantic decomposition trees*

(a) Use a semantic decomposition tree to show that $(p \wedge q) \rightarrow r$ |- $(p \wedge (q \vee s)) \rightarrow r$.

(b) Use semantic decomposition trees to determine whether each the following hold: (i) $(p \vee q) \rightarrow r$ -||- $(p \rightarrow r) \wedge (q \rightarrow r)$, (ii) $p \rightarrow (q \wedge r)$ -||- $(p \rightarrow q) \wedge (p \rightarrow r)$, (iii) $(p \wedge q) \rightarrow r$ -||- $(p \rightarrow r) \vee (q \rightarrow r)$, (iv) $p \rightarrow (q \vee r)$ -||- $(p \rightarrow q) \vee (p \rightarrow r)$.

(c) Use semantic decomposition trees to determine whether $p \rightarrow (q \rightarrow r)$ |- $(p \rightarrow q) \rightarrow r$ and conversely.

8.6. *Natural deduction*

(a) Show that $\{p \rightarrow ((r \vee s) \rightarrow t), q \rightarrow (\neg(s \vee u) \vee t), s\}$ |- $(p \vee q) \rightarrow t$ by constructing a natural deduction that uses conditional proof followed by disjunctive proof. Pay careful attention to the 'book-keeping' columns!

(b) Do the same, but using conditional proof followed by proof by contradiction.

Selected Reading

Introductions to discrete mathematics tend to put their chapters on logic right at the beginning. In the order of nature, this makes sense, but it makes it difficult to use helpful tools like set, relation and function in the exposition. One text that introduces logic after having presented those notions is:

James Hein *Discrete Structures, Logic and Computability*. Jones and Bartlett, 2002 (second edition), Chapter 6.

Four well-known books on elementary logic are listed below. The first is written specifically for students of computer science without much mathematics, the second for the same but with more mathematical sophistication, while the last two are aimed at students of philosophy and the general reader.

Michael Huth and Mark Ryan *Logic in Computer Science*. Cambridge University Press, 2000, Chapter 1.

Mordechai Ben-Ami *Mathematical Logic for Computer Science*. Springer, 2001 (second edition), Chapters 1–4.

Colin Howson *Logic with Trees*. Routledge, 1997, Chapters 1–4.

Wilfrid Hodges *Logic*. Penguin, 1977, Sections 1–25.1.

9

Something about Everything: Quantificational Logic

Chapter Outline

Although fundamental, the logic of truth-functional connectives has very limited expressive power. In this chapter we go further, explaining the basic concepts of *quantificational* (alias *first-order*, or again *predicate*) logic, which is sufficiently expressive to cover all of the deductive reasoning that is carried out in standard mathematics.

We begin by presenting the *language* of quantificational logic – its components such as the *universal and existential quantifiers*, and the ways they can be put to work to express complex relationships. With only an intuitive understanding of the quantifiers, some of the basic *logical equivalences* involving them can already be appreciated. Although the language still works within the assumption of bivalence, truth-tables are not sufficient to give an analysis of the two quantifiers. For this reason, we need to provide a rather more complex *semantics*. This is followed by a review of some of the most important *logical implications* with quantifiers. These provide the basis for extending the procedures of *natural deduction* to deal with the quantifiers.

9.1 The Language of Quantifiers

We have been using quantifiers informally throughout the book, and have made a few remarks on them in logic boxes. Recall that there are two quantifiers ∀ and ∃, meaning 'for all' and 'for some'. They are always used with an attached variable.

D. Makinson, *Sets, Logic and Maths for Computing*,
DOI: 10.1007/978-1-84628-845-6_9, © Springer-Verlag London Limited 2008

9.1.1 Some Examples

Before getting systematic, we give some examples of statements of ordinary English and their representation using quantifiers, comment on their salient features, and raise questions that will be answered as we continue the chapter

Table 9.1 Examples of quantified statements

	English	Symbols
1	All composer are poets	$\forall x(Cx \rightarrow Px)$
2	Some composers are poets	$\exists x(Cx \wedge Px)$
3	No poets are composers	$\forall x(Px \rightarrow \neg Cx)$
4	Everybody loves someone	$\forall x \exists y(Lxy)$
5	There is someone who is loved by everyone	$\exists y \forall x(Lxy)$
6	There is a prime number less than 5	$\exists x(Px \wedge (x{<}5))$
7	Behind every successful man stands an ambitious woman	$\forall x((Mx \wedge Sx) \rightarrow \exists y(Wy \wedge Ay \wedge Byx))$
8	No man is older than his father	$\neg \exists x(Oxf(x))$
9	The successors of distinct integers are distinct	$\forall x \forall y(\neg(x \equiv y) \rightarrow \neg(s(x) \equiv s(y)))$

	Comments	Questions
1	Uses the universal quantifier $\forall$, variable x, two predicate letters, truth-functional $\rightarrow$	Could we use a different variable, say y? Why are we using $\rightarrow$ here instead of, say, $\wedge$? Can we express this using $\exists$ instead of $\forall$? What is its relation to $\forall x(Px \rightarrow Cx)$?
2	Uses the existential quantifier $\exists$, truth-functional $\wedge$	Why are we using $\wedge$ here instead of $\rightarrow$? Can we express this using $\forall$ instead of $\exists$? Is it logically implied by 1? What is its relation to $\exists x(Cx \wedge \neg Px)$?
3	Uses $\forall, \rightarrow, \neg$	Can we express this using $\exists, \wedge, \neg$? What is its relation to $\forall x(Cx \rightarrow \neg Px)$?
4	Two quantifiers, relation symbol	Why haven't we used a predicate P for 'is a person'? Does the meaning change if we write $\forall x \exists y(Lyx)$?
5	Order of quantifiers-with-attached-variables reversed	Is this equivalent to 4? Does either logically imply the other?
6	Uses a constant '5'	Could we somehow replace the individual constant by a predicate?
7	Two quantifiers, more complex truth-functional part	Could we express this equivalently with both quantifiers up the front?
8	Uses a function symbol f with $f(x)$ meaning 'the father of x'	Could we express this with a relation symbol F and two variables?
		Could we express the statement more 'positively'?
9	No quantifier explicit in the English, but two universals are implicit. Uses the identity relation symbol, which we write as $\equiv$ rather than $=$ to avoid confusions when we define the semantics shortly.	Does the meaning change if we reverse the initial quantifiers? Is the identity relation any different, from the point of view of logic, than other relations?

EXERCISE 9.1.1

On the basis of your informal experience with quantifiers, have a go at answering the questions in the table. If you feel confident about most of them, congratulations, you seem to have a head start! But keep a record of your answers to check later. If the questions leave you puzzled or uncertain, by the end of this chapter you should be able to do better.

9.1.2 Systematic Presentation of the Language

The basic components of the language of quantificational logic are set out in the following table.

Table 9.2 Ingredients of the language of quantificational logic

Broad Category	Specific Items	Signs Used	Purpose
Designators	constants	$a, b, c, \ldots$	name specific objects: e.g. 5, Charlie Chaplin, London
	variables	$x, y, z, \ldots$	range over specific objects, combine with quantifiers to express generality
Function letters	1-place n-place	$f, g, h, \ldots$	form compound terms out of simpler terms
Predicates	1-place	$P, Q, \ldots$	e.g. is prime, is funny, is polluted
	2-place	$P, Q, R, S, \ldots$	e.g. is smaller than, resembles, is colder than
	n-place	$P, Q, R, S, \ldots$	e.g. lies between (3-place)
	special relation sign	$\equiv$	identity
Quantifiers	universal	$\forall$	for all
	existential	$\exists$	there is
Connectives	from propositional logic	$\neg, \wedge, \vee, \rightarrow,$ etc	usual truth-tables
Auxiliary	parentheses and commas	$(,), , ,$	parentheses are necessary to ensure unique decomposition, commas are optional to make formula easier to read.

In each of the categories of designators, function letters and predicates, we assume that we have an infinite supply. These can be referred to by using numerical subscripts, e.g. $f_1, f_2, \ldots$ for the function letters. Strictly speaking, we should also have numerical superscripts to indicate the arity of the function and predicate letters, so that the two-place predicate letters, say, are listed as

$R^2{}_1$, $R^2{}_2$,... But as this is rather cumbersome, we ordinarily use a few chosen letters as indicated in the table, with context or comment indicating the arity.

How are these ingredients put together to build formulae? Recursively, as you would expect, but in two stages. First we define the notion of a *term*.

> *Basis* : Constants and variables are terms.
> *Recursion step* : If f is an n-place function symbol and t_1, ... t_n are terms, then $f(t_1, \ldots , t_n)$ is a term.

As remarked above, commas are not really needed, but are customarily used to make terms easier to read. Moreover, as all function symbols are prefixed to their arguments, and each function symbol has a predetermined arity, we do not really need parentheses when forming terms (recall the discussion of parenthesis-free notation in the chapter on trees). But parentheses are usually included to facilitate human reading, as in the above definition.

EXERCISE 9.1.2 (WITH SOLUTION)

> Which of the following are terms, where f is 2-place and g is 1-place? In the negative cases, give a brief reason. (i) a, (ii) ax, (iii) $f(x,b)$, (iv) $f(b,g(y))$, (v) $g(g(x,y))$, (vi) $f(f(b,b), f(a,y))$, (vii) $g(g(a))$, (viii) $g(g(a)$, (ix) $gg(a)$.
>
> *Solution:* (i) Yes. (ii) No: it is neither a constant nor a variable nor formed using a function symbol. (iii) Yes. (iv) Yes. (v) No: g is one place. (vi) Yes. (vii) Yes, (viii) No: a right-hand parenthesis forgotten. (ix) Strictly speaking no, but this often used to abbreviate $g(g(a))$, and we will do the same in what follows.

Given the terms, we can now define formulae recursively.

Basis: If R is an n-place predicate and t_1,...,t_n are terms, then $R(t_1,\ldots,t_n)$ is a *formula* (called an *atomic formula*), and so is $(t_1 \equiv t_2)$ where $\equiv$ is the symbol for the identity relation.

Recursion step: If α, β are formulae and x is a variable, then the following are formulae: $(\neg\alpha)$, $(\alpha\wedge\beta)$, $(\alpha\vee\beta)$, $(\alpha\rightarrow\beta)$, $\forall x(\alpha)$, $\exists x(\alpha)$.

As in propositional logic we drop parentheses whenever context or convention suffices to ensure unique decomposition. For ease of reading, when there are multiple parentheses, we may also use different styles, e.g. square brackets. The special relation sign $\equiv$ for identity is customarily infixed. All other predicates, like function symbols, are prefixed; for this reason the parentheses and commas in atomic formulae without identity can be omitted without ambiguity, and sometimes will be.

Alice Box: Metavariables, use and mention

Alice: In the recursion step, where you consider the quantifiers, you have used one particular variable, namely x. Do you mean that we can use any of the variables x, y, z,... in this position?

Hatter: Of course. So when α is a formula, then $\forall y(\alpha), \forall z(\alpha)$ etc are also formulae.

Alice: So why didn't you say so explicitly?

Hatter: We could have done so, but to do it systematically we would have to draw on a fresh supply of *metavariables*. These would be variables in our *metalanguage* (i.e. the language in which we are presenting quantificational logic), and would range over the set of all signs serving as variables in the *object language* (in this instance the language of quantificational logic). Indeed, we would have to do the same when talking about predicates, function symbols etc. In the early twentieth century, as part of a desire to make everything absolutely explicit, there were a few books that did all this, whether with a whole array of such metavariables or with the help of a special device known as *quasi-quotes*. But you can imagine how cumbersome this is ! The text quickly becomes unreadable and intensely annoying. So nobody does it any more, although everybody knows that it *can* be done, and indeed for utter strictness should be done.

Alice: I notice that when you talk about a specific sign, say the conjunction sign or the universal quantifier, you do not put it in inverted commas. You say: 'The symbol $\forall$ is the universal quantifier' rather than 'The symbol "$\forall$" is the universal quantifier'. Is this legitimate? Isn't it like saying 'The word London has six letters' when you should be saying 'The word "London" has six letters'?

Hatter: Indeed, strictly speaking we should use some device like quotation marks, italicizing, or underlining to distinguish use and mention – the *use* of a symbol like $\forall$ in our object language and its *mention* in our metalanguage. But this is also very tiresome when done systematically, and we can usually omit it without confusion.

Alice: So presentations of logic are in general rather less rigorous than the logic they present!

Hatter: I suppose you could put it that way...

EXERCISE 9.1.3 (WITH SOLUTION AND COMMENTS)

(a) Which of the following are formulae, where R is a 2-place predicate and P is a 1-place predicate, and f, g are as in the previous exercise? In the negative cases, give a brief reason. (i) $R(a,a)$, (ii) $R(x,b) \wedge \neg P(g(y))$,

(iii) $R(x,P(b))$, (iv) $P(P(x))$, (v) $\forall(Rxy)$, (vi) $\exists P(Px)$, (vii) $\forall x\forall x(Px)$, (viii) $\forall x(Px{\rightarrow}Rxy)$, (ix) $\exists y\exists y(R(f(x,y),g(x)))$, (x) $\forall x(\exists y(Rxy))$, (xi) $\forall x\exists y(Rxy)$, (xii) $\exists x(Ray)$, (xiii) $\exists x\forall x(Ray)$, (xiv) $\exists x\exists y\forall z\forall w\exists u(P(x)\wedge R(y,z)\wedge R(w,u))$.

(b) Go through all the symbolic expressions in Table 9.1 and check that they are really formulae in the sense defined (given conventions for the omission of brackets such as we have described).

Solution and comments:

(a) (i), (ii) Yes. (iii) No: $P(b)$ is a *formula*, not a *term*, and we need a term in this position if the whole expression is to be a formula. Note this carefully, as it is a *common student error*. (iv) No, same reason. (v) No: the quantifier does not have any variable attached to it. (vi) No: the quantifier has a predicate attached to it, rather than a variable. Such expressions are admitted in what is called *second-order logic*, but not in quantificational (alias first-order) logic. (vii) Yes, despite the fact that both quantifiers have the same variable attached to them. (viii), (ix), (x) Yes. (xi) Strictly speaking, a pair of parentheses is missing. But in this kind of formula, where we have a string of quantifiers following each other, we can drop such parentheses without ambiguity, and will do so. (xii), (xiii) Yes. (xiv) Yes, although the mind boggles at this sequence of five initial quantifiers (best thought of as three blocks of alternating quantifiers). Such levels of quantificational complexity do actually occur in working mathematics: for an example of three alternating quantifiers, think of the definition of a limit, if you are familiar with it.

(b) Yes, they are all formulae, given those conventions. Notice the omission of parentheses in the double quantifications 4 and 5, and the infixing of the identity sign in 9 (and of the sign for an arithmetic relation in 6).

EXERCISE 9.1.4 (WITH SOME ANSWERS, HINTS AND COMMENTS)

Express the following statements in the language of quantificational logic, using naturally suggestive letters for the predicates. For example, in (a) use L for the predicate 'is a lion', T for the predicate 'is a tiger', and D for the predicate 'is dangerous'.

(a) Lions and tigers are dangerous, (b) If a triangle is right-angled then it is not equilateral, (c) Anyone who likes Albert likes Betty, (d) Albert doesn't like everybody Betty likes, (e) Albert doesn't like anybody Betty likes, (f) Everyone who loves someone is loved by that person, (g) Everyone who loves someone is loved by someone, (h) There are exactly two prime numbers less than 5, (i) Every integer has exactly one successor (j) My mother's father is older than my father's mother.

Some answers, hints and comments:

(a) Can be $\forall x(Lx{\to}Dx)\wedge\forall x(Tx{\to}Dx)$ or $\forall x((Lx\vee Tx){\to}Dx)$. It cannot be expressed as $\forall x((Lx\wedge Tx){\to}Dx)$ – try to understand why! Notice that the universal quantifier is not explicit in the English, but it evidently part of the meaning.

(b) $\forall x((Tx\wedge Rx){\to}\neg Ex)$. Here we are using T for 'is a triangle'. You can also simplify life by declaring the set of all triangles as your 'domain of discourse' and write simply $\forall x(Rx{\to}\neg Ex)$. This kind of simplification, by declaring a domain of discourse, is often useful when symbolizing. But it is legitimate only if the domain is chosen so that nothing the statement says concerns anything outside the domain.

(c) $\forall x(Lxa{\to}Lxb)$. We declare our domain of discourse to be the set of all people, write Lxy stands for 'x likes y', and we use individual constants for Albert and Betty.

(d), (e) *Hint*: Try to understand the difference of meaning between the two. The English words 'every' and 'any' tend do much the same work when they occur positively, but when occurring negatively they do quite different jobs. Declare your domain of discourse.

(f), (g) *Hint*: Try to understand the difference of meaning, and declare your domain of discourse. In (f) you will need two quantifiers, both universal (despite the idiomatic English), while in (g) you will need three quantifiers.

(h) *Hint*: The problem here is how to say that there are *exactly two* items with a certain property. Try paraphrasing it as: there is an x and there is a y such that they both have the property and everything with the property is identical to at least one of them. This goes into the language of quantificational logic smoothly. Take your domain of discourse to be the set of positive integers.

(i) *Hint*: If you can express 'exactly two' it should be easy to express 'exactly one'.

(j) *Hint*: Declare your domain of discourse. Use one-place function letters for 'father of' and 'mother of'. Use a constant to stand for yourself.

9.1.3 Freedom and Bondage

To understand the internal structure of a formula of quantificational logic, three notions are essential – the *scope* of a quantifier, and *free* versus *bound* occurrences of a variable. We explain them through some examples.

Consider the formula $\forall z(\neg Rxz \wedge \exists y(Rxy))$. It has two quantifiers. The *scope* of the each quantifier is the material in the parentheses immediately following it. Thus the scope of the first quantifier is the material between the large square brackets:

$$\forall z[\neg Rxz \wedge \exists y(Rxy)].$$

The scope of the second quantifier likewise:

$$\forall z(\neg Rxz \wedge \exists y[Rxy]),$$

Note how the scope of one quantifier may lie inside the scope of another, or be entirely separate from it. They never overlap.

In a quantified formula $\forall x(\alpha)$ or $\exists x(\alpha)$ the quantifier is said to *bind* the occurrence of the variable x that is attached to it, and all occurrences of the same variable x occurring in α, unless some other quantifier occurring inside α already binds them. An occurrence of a variable x in a formula α is said to be *bound in* α iff there is some quantifier in α that binds it. Occurrences of a variable that are not bound in a formula are called *free* in the formula. Finally, a formula with no free occurrences of any variables is said to be *closed*.

EXERCISE 9.1.5 (WITH PARTIAL SOLUTION)

(a) Identify the free and bound occurrences of variables in the formula $\forall z(\neg Rxz \wedge \exists y(Rxy))$. Use arrows from above to mark the bound occurrences, arrows from below for the free ones.

(b) Use large square brackets to indicate the scope of each quantifier in the formula $\forall x \exists y(Rxy) \vee \exists z(Py \rightarrow \forall x(Rzx \wedge Ryx))$. For this, write the formula out four times, once for each quantifier. Recall the convention for omitting parentheses in a string of contiguous quantifiers.

(c) Identify the bound and the free occurrences of variables in the formula of (b), using arrow pointers as before.

(d) Use the last example to illustrate how a single variable may have one occurrence bound, another free, in the same formula.

(e) Give a simple example to illustrate how an occurrence of a variable may be bound in a formula but free in one of its subformulae.

(f) Consider the formula $\forall x \exists y \forall x \exists y (Rxy)$. Which quantifiers bind which occurrences of x? Which bind which occurrences of y?

(g) Which of the formulae mentioned in this Exercise are closed?

Solution to (f): The outer $\forall$ binds *only* the occurrence of x attached to it. The inner $\forall$ binds *both* the occurrence of x attached to it and the occurrence of x in Rxy. Likewise for $\exists$.

Comment: The translation into symbols of any complete sentence of English should have all its variables bound, i.e. it should be closed. Review Table 9.1 to check this out.

9.2 Some Basic Logical Equivalences

At this point we could set out the semantics for quantificational formulae, with rigorous definitions of logical relationships such as consequence and equivalence. But we will postpone the formal definitions until the next section, and in the meantime cultivate intuitions. There are a number of basic logical equivalences that can already be appreciated when you read $\forall x(\alpha)$ and $\exists x(\alpha)$ intuitively as saying respectively 'α holds for every x in our domain of discourse' and 'α holds for at least one x in our domain of discourse'.

First among these equivalences are the *quantifier interchange* principles, which show that anything expressed by one of our two quantifiers may equivalently be expressed by the other with the judicious help of negation. In the following table, the formulae on the left are logically equivalent to those on the right.

Table 9.3 Quantifier interchange equivalences

$\neg\forall x(\alpha)$	$\exists x(\neg\alpha)$
$\neg\exists x(\alpha)$	$\forall x(\neg\alpha)$
$\forall x(\alpha)$	$\neg\exists x(\neg\alpha)$
$\exists x(\alpha)$	$\neg\forall x(\neg\alpha)$

Here α stands for any formula of quantificational logic. We will use -‖- for the as yet not formally defined, but intuitively understood, concept of logical equivalence for quantificational formulae.

Actually, any one of these four equivalences can be obtained from any other by means of double negation and the principle of replacement of logically equivalent formulae. For example, suppose we have the first one, and we want to get the last one. We note that $\exists x(\alpha)$ -‖- $\exists x(\neg\neg\alpha)$ -‖- $\neg\forall x(\neg\alpha)$ and we are done.

EXERCISE 9.2.1

(a) Obtain the second and third quantifier interchange principles from the first one by a similar procedure.

(b) Use quantifier interchange and suitable truth-functional equivalences to show that (i) $\neg\forall x(\alpha\rightarrow\beta)$ -‖- $\exists x(\alpha\wedge\neg\beta)$, (ii) $\neg\exists x(\alpha\wedge\beta)$ -‖- $\forall x(\alpha\rightarrow\neg\beta)$, (iii) $\forall x(\alpha\rightarrow\beta)$ -‖- $\neg\exists x(\alpha\wedge\neg\beta)$, (iv) $\exists x(\alpha\wedge\beta)$ -‖- $\neg\forall x(\alpha\rightarrow\neg\beta)$. These equivalences are important, because the formulae correspond to familiar kinds of statement in English. For example in (i), the left side says 'not all αs are βs', while the right one says 'at least one α is not a β'.

(c) To what English statements do the equivalences (ii) through (iv) correspond?

The next group of equivalences may be described as *distribution principles*. They show that way in which universal quantification distributes over conjunction, while the existential distributes over disjunction.

Table 9.4 Distribution equivalences

$\forall x(\alpha\wedge\beta)$	$\forall x(\alpha)\wedge\forall x(\beta)$
$\exists x(\alpha\vee\beta)$	$\exists x(\alpha)\vee\exists x(\beta)$

Why does the universal quantifier get on so well with conjunction? Essentially because it is like a long conjunction. Suppose our domain of discourse has just n elements, named by n individual constants $a_1,\ldots,a_n$. Then saying $\forall x(Px)$ amounts to saying, of each element in the domain, that *it* has the property, i.e. that $Pa_1\wedge\ldots\wedge Pa_n$. This is called a *finite transform* of $\forall x(Px)$.

Of course if we change the size of the domain, then we change the length of the conjunction; and to make the breakdown work we must have enough constants to name all the elements of the chosen domain. But even with these provisos, it is clear that the principles for $\forall$ tend to reflect those for $\wedge$, while those for $\exists$ resemble familiar ones for $\vee$. We will return to this after giving the semantics for quantificational logic.

EXERCISE 9.2.2

(a) Write out $\forall x(\alpha \wedge \beta)$ as a conjunction in a domain of three elements, with α taken to be the atomic formula Px and β the atomic formula Qx. Then write out $\forall x(\alpha) \wedge \forall x(\beta)$ in the same way and check that they are equivalent. What truth-functional principles do you appeal to in that check?

(b) What does $\exists x(Px)$ amount to in a domain of n elements?

(c) Write out $\exists x(\alpha \vee \beta)$ as a disjunction in a domain of three elements, with α chosen as Px and β as Qx. Then write out $\exists x(\alpha) \vee \exists x(\beta)$ in the same way and check that they are equivalent. What truth-functional principles do you appeal to?

(d) If we think of the universal and existential quantifier as expressing generalized conjunctions and disjunctions respectively, to what familiar truth-functional equivalences do the entries in Table 9.3. correspond?

(e) Illustrate your answer to (d) by constructing finite transforms for the third row of Table 9.3. in a domain of three elements with α chosen as Px.

This idea of working with such finite transforms of quantified formulae can be put to systematic use to obtain negative results. We will return to this later.

The last group of intuitively obvious equivalences that we will mention are the vacuity principles. They are expressed in the following table, where all formulae in a given row are equivalent.

Table 9.5 Vacuity equivalences

$\forall x(\alpha)$	$\forall x \forall x(\alpha)$	$\exists x \forall x(\alpha)$
$\exists x(\alpha)$	$\exists x \exists x(\alpha)$	$\forall x \exists x(\alpha)$

In other words, quantifying twice on the same variable does no more than doing it once. To say 'for every x it is true that for every x we have Px' is just a longwinded way of saying 'Px for every x'. Note that in these vacuity equivalences *the variables must be the same*: $\forall x(Rxy)$ is *not* logically equivalent to $\forall y \forall x(Rxy)$ or to $\exists y \forall x(Rxy)$.

The equivalences of the table are special cases of a general *principle of vacuous quantification*: If there are no free occurrences of x in φ, then each of $\forall x(\varphi)$ and $\exists x(\varphi)$ is logically equivalent to φ.

9.3 Semantics for Quantificational Logic

It is time to get rigorous. In this section we present the semantics for quantificational logic. To keep a sense of direction, remember that we are working towards an analogue, adequate for our enriched language, of the concept of a valuation of propositional logic and the ensuing notions of consequence etc.

9.3.1 Interpretations

We begin with the notion of an *interpretation*. This is defined to be any pair (D,δ) where:

- D is a non-empty set (called the *domain* or *universe* of discourse of the interpretation)

- δ is a function (called the *designation* or *denotation* function) that assigns a value to each constant, variable, predicate letter and function letter of the language in the following way:

  - For each constant a of the language, $\delta(a)$ is an element of the domain, i.e. $\delta(f) \in D$,

  - Likewise, for each variable x of the language, $\delta(x)$ is an element of the domain, i.e. $\delta(x) \in D$,

  - For each n-place function letter f of the language, $\delta(a)$ is an n-argument function on the domain, i.e. $\delta(f): D^n \to D$,

  - For each n-place predicate P of the language, other than the identity symbol, $\delta(P)$ is an n-place relation over the domain, i.e. $\delta(P) \subseteq D^n$.

  - For the identity symbol, we put $\delta(\equiv)$ to be the identity relation over the domain.

The last clause of this definition means that we are giving the identity symbol privileged treatment. Whereas other predicate symbols may be interpreted as any relations over the domain, so long as they have the right arity, the identity symbol is always treated as *genuine identity* over the domain. Semantics following this rule are often said to be *standard* (with respect to identity). It is also possible to give a non-standard treatment of identity, in which it is interpreted as any equivalence relation over the domain that is also a congruence with respect to all function letters and predicates. We will not go further into the non-standard approach, confining ourselves to standard interpretations.

9.3.2 Valuating Terms Under an Interpretation

Given an interpretation (D,δ), we can define the value $\delta^+(t)$ of a term recursively, following the recursive definition of the terms themselves:

$$\delta^+(a) = \delta(a) \text{ for every constant } a \text{ of the language}$$
$$\delta^+(x) = \delta(x) \text{ for every variable } x \text{ of the language}$$
$$\delta^+(f(t_1, \ldots, t_n)) = (\delta^+(f))(\delta^+(t_1), \ldots, \delta^+(t_n)).$$

The recursion step (i.e. the last clause) needs some commentary. We are in effect defining $\delta^+(f(t_1, \ldots, t_n))$ homomorphically. The value of a term $f(t_1, \ldots, t_n)$ is defined by taking the interpretation of the function letter f, which will be a function on D^n into D, and applying that function to the interpretations of the terms $t_1, \ldots, t_n$, which will all be elements of D, to get a value that will also be an element of D. The notation makes it look complicated, but the underlying idea is natural and simple.

There are some contexts in which the notation could be simplified. When considering only *one* interpretation function δ, we can reduce clutter by writing $\delta(t)$ as $\underline{t}$, simply underlining the term t. Written this way, the recursion clause of the above definition says: $\underline{f(t_1, \ldots, t_n)} = f(\underline{t_1}, \ldots, \underline{t_n})$ – which is much easier to read. But in contexts where we are considering more than one interpretation (which is most of the time), this simple notation is not open to us, unless of course we use more than one kind of underlining.

As δ^+ is a uniquely determined extension of δ, we will follow the same 'abuse of notation' as in propositional logic and write it simply as δ.

9.3.3 Valuating Formulae Under an Interpretation: Basis

Finally, we may go on to define the notion of the *truth* or *falsehood* of a formula α under an interpretation (D,δ). Again our definition is recursive, following the recursive construction of the formulae themselves. We are defining a function ν from the set L of all formulae into the set $\{0,1\}$, i.e. $\nu: L \rightarrow \{0,1\}$. So we will merely specify when $\nu(\alpha) = 1$, leaving it understood that otherwise $\nu(\alpha) = 0$. Of course, strictly speaking ν should be written as $\nu_{D,\delta}(\alpha)$, but in our battle to keep notation under control we omit the subscript whenever context allows us to do so without serious ambiguity.

- *Basis, first part:* For any atomic formula of the form $Pt_1, \ldots, t_n$: $\nu(Pt_1, \ldots, t_n) = 1$ iff $(\delta(t_1), \ldots, \delta(t_n)) \in \delta(P)$. In the shorthand underlining notation: $P\underline{t_1}, \ldots, \underline{t_n} = 1$ iff $(\underline{t_1}, \ldots, \underline{t_n}) \in \underline{P}$.

- *Basis, second part:* For any atomic formula of the form $t_1 \equiv t_2$: $\nu(t_1 \equiv t_2) = 1$ iff $(\delta(t_1), \delta(t_2)) \in \delta(\equiv)$. Since $\delta(\equiv)$ is required to be the identity relation over D, this means that $\nu(t_1 \equiv t_2) = 1$ iff $\delta(t_1) = \delta(t_2)$.

9.3.4 Valuating Formulae Under an Interpretation: Recursion Step

The recursion step also has two parts, one for the truth-functional connectives and the other for the quantifiers. The first part is easy.

- *Recursion step for the truth-functional connectives*: $v(\neg\alpha) = 1$ iff $v(\alpha) = 0$ and so on for the other truth-functional connectives, just as in propositional logic.

The *recursion step for the quantifiers* is at the heart of quantificational logic. But it is subtle, and needs careful formulation. In the literature, there are two ways of doing this. They are equivalent, but passions can run high over which is better to use, and it is good policy to understand both of them. They are known as the *x-variant* and *substitutional* readings of the quantifiers.

9.3.5 The x-Variant Reading of the Quantifiers

Let (D,δ) be any interpretation, and let x be any variable of the language. By an *x-variant* of (D,δ) we mean any interpretation (D,δ') that agrees with (D,δ) in the choice of domain, and also in the interpretation given to all constants, function letters and predicates (i.e. $\delta'(a) = \delta(a)$, $\delta'(f) = \delta(f)$, $\delta'(P) = \delta(P)$ for all letters of the respective kinds), and also agrees on the interpretation given to all variables *except possibly the variable x* (so that $\delta'(y) = \delta(y)$ for every variable y of the language other than the variable x). With this notion in hand, we evaluate the quantifiers as follows:

$\nu(\forall x(\alpha)) = 1$ where $\nu = \nu_{D,\delta}$, iff $\nu_{D,\delta'(\alpha)} = 1$ for every x-variant interpretation (D, δ') of (D, δ).

$\nu(\exists x(\alpha)) = 1$ where $\nu = \nu_{D,\delta}$, iff $= \nu_{D,\delta'\,(\alpha)} = 1$ for at least one x-variant interpretation (D, δ') of (D, δ).

Alice Box: The x-variant reading of the quantifiers

Alice: So we are using quantifiers in our metalanguage when defining their semantics in the object language?

Hatter: Sure, just as we used truth-functional connectives in the metalanguage when defining their semantics in the object language. There's no other way.

EXERCISE 9.3.1 (WITH PARTIAL SOLUTION)

(a) In a domain D consisting of two items 1 and 2, construct an interpretation that makes the formula $\forall x(Px \vee Qx)$ true but makes $\forall x(Px) \vee \forall x(Qx)$ false. Show your calculations of the respective truth-values.

(b) Do the same for the pair $\exists x(Px) \wedge \exists x(Qx)$ and $\exists x(Px \wedge Qx)$.

(c) Again for the pair $\forall x \exists y(Rxy)$ and $\exists y \forall x(Rxy)$.

Sample solution to (a): Put $\delta(P) = \{1\}$, $\delta(Q) = \{2\}$, and $\delta(x)$ any element of D. Let $v = v_{D,\delta}$. We claim that $v(\forall x(Px \vee Qx)) = 1$ while $v(\forall x(Px) \vee \forall x(Qx)) = 0$. For the former, it suffices to show that $v'(Px \vee Qx) = 1$ for every x-variant interpretation v'. But there are only two such x-variants: one puts $\delta'(x) = 1$ while the other puts $\delta''(x) = 2$. The former gives us $v'(Px) = 1$ while the latter gives $v''(Qx) = 1$, so $v'(Px \vee Qx) = v''(Px \vee Qx) = 1$ and thus $v(\forall x(Px \vee Qx)) = 1$. For the latter, we have $v''(Px) = 0 = v'(Qx)$ so that $v(\forall x(Px)) = 0 = v(\forall x(Qx))$ and so finally $v(\forall x(Px) \vee \forall x(Qx)) = 0$ as desired.

Comments: Evidently, a detailed verification like this is rather tedious, and before long you should be able to 'see' the value of a fairly simple formula under a given interpretation without writing out all the details. But to get to that point, one has to begin by doing some examples in full detail.

After working through these examples, it should be clear that the truth-value of a formula under an interpretation is independent of the value given to the variables that do not occur free in it. For example, in the formula $\forall x(Px \vee Qx)$ the variable x does not occur free – all its occurrences are bound – and so its truth-value under the interpretation (D, δ) is independent of the value of $\delta(x)$.

Alice Box: Interpreting formulae that are not closed

Alice: This may connect with something that bothers me. I have been browsing in several books on logic for students of mathematics. They use the x-variant account, but in a rather different way. Under the definition above, every formula comes out as *true*, or as *false*, under a given interpretation, i.e. we always have either $v_{D,\delta}(\alpha) = 1$ or $v_{D,\delta}(\alpha) = 0$. But the books that I have been looking at insist that when a formula α has free occurrences of variables (i.e. is not closed) then in general it is *neither true nor false under an interpretation*! What is going on?

(Continued)

Alice Box: (Continued)

Hatter: Indeed, you are quite right to be puzzled! That is another way of building the same x-variant semantics. When those authors speak of an 'interpretation' or *model*, they mean one which is like ours except that the designation function δ does not act on the variables of the language – only on the constants, function letters, and predicates. A model, in that sense, is then supplemented by an *assignment* of values in D to the variables. Thus our interpretation is their *model-plus-assignment*.

Alice: And then?

Hatter: In such texts, formulae (including those with free occurrences of variables) are evaluated under a model-plus-assignment in exactly the same way as we have done under an interpretation, but with a terminological difference. Rather than speaking of truth and falsity and writing $v_{D,\delta}(\alpha) = 1$ and $v_{D,\delta}(\alpha) = 0$, they speak of the model-plus-assignment 'satisfying' or 'not satisfying' the formula. Finally, a formula is declared to be *true in the model* iff it is satisfied for *every assignment* accompanying that model, and is called *false in the model* iff it is satisfied for *none* of those assignments. Those formulae that are satisfied by some but not all of the assignments accompanying the model (for example the simple atomic formula Px, when the model interprets the predicate letter P as a proper non-empty subset of the domain) are deemed to be neither true nor false in the model. As a result, an arbitrary formula α is counted as true-in-a-model iff $\forall x(\alpha)$ is true in it.

Alice: Why not follow that way of speaking?

Hatter: No objection in principle: the formulations are equally rigorous, and give the same result for closed formulae. For formulae that are not closed, our 'truth under an interpretation' is the same as their 'satisfaction under a model-plus-assignment'. But my experience is that such a way of speaking causes unnecessary muddles for students.

Alice: How?

Hatter: Having become accustomed to expressing bivalence in propositional logic with the terms 'truth' and 'falsehood', when they get to quantificational logic they are asked to express the same fundamental principle with the terms 'satisfaction' and 'non-satisfaction', and to employ the terms 'truth' and 'falsehood' in a non-bivalent way. A sure recipe for classroom confusion!

9.3.6 The Substitutional Reading of the Quantifiers

The second way of reading the quantifiers is not often used by mathematicians, with some exceptions – for example in celebrated work of Abraham Robinson. But it is quite often employed by computer scientists and philosophers.

Given a formula α and a variable x, we write $\alpha[t/x]$ to stand for the result of *substituting the term t for all free occurrences of x in* α. Thus if α is the formula $\forall z(\neg Rxz \wedge \exists x(Rxy))$ then $\alpha[a/x] = \forall z(\neg Raz \wedge \exists x(Rxy))$, obtained by replacing the unique free occurrence of x in α by the constant a.

EXERCISE 9.3.2

    (a) Let α be the formula $\forall z(\neg Rxz \wedge \exists y(Rxy))$ used earlier when defining the notion of scope. Write out $\alpha[a/x]$, $\alpha[b/y]$.

    (b) Do the same for the formula $\forall x \exists y(Rxy) \vee \exists z(Py \rightarrow \forall x(Rzx \wedge Ryx))$, also introduced in Exercise 9.1.5.

We are now ready to give the substitutional reading of the quantifiers. Let (D, δ) be any interpretation. Make sure that the interpretation function δ, restricted to the set of all constants of the language, is *onto* D, i.e. that every element of D is the value under δ of some constant symbol. If it is not already onto D, add enough constants to the language, extending the domain of the interpretation function to ensure that it is onto. Then evaluate the quantifiers as follows:

$\nu(\forall x(\alpha)) = 1$ iff $\nu(\alpha[a/x]) = 1$ for every constant symbol
    a of the (thus expanded) language.

$\nu(\exists x(\alpha)) = 1$ iff $\nu(\alpha[a/x]) = 1$ for at least one constant symbol
    a of the (thus expanded) language.

Alice Box: Substitutional reading of the quantifiers

Alice: One moment! There is something funny here. I see why you need to expand the supply of constants in the language. You need to guarantee that $\forall x$ means 'for every element in the domain' and not just 'for every element of the domain that happened to get a name'. But this means that the language is

(Continued)

Alice Box: (Continued)

not fixed – the supply of constants depends on the particular domain of discourse under consideration. That's odd!

Hatter: Odd, yes, but not incoherent. It works perfectly well. It is not difficult (though rather messy and tedious) to show that every formula α in the original, unexpanded language gets the value 1 under the x-variant reading of the quantifiers iff after expanding the language, it gets the value 1 under the substitutional reading.

Alice: Still, I would like to keep my language fixed.

Hatter: Well, up to a point we can do that by making use of a deep result known as the *Löwenheim-Skolem theorem*. Perhaps you remember, from one of our conversations in the chapter on functions, the concept of a *countable set*: it is one that has a bijection (one-one correspondence) with the set **N** of natural numbers. The Löwenheim-Skolem theorem tells us that for any formula α (indeed, for any set A of formulae) of quantificational logic, if there is an interpretation (D,δ) with $v_{(D,\delta)}(\alpha) = 1$, then there is an interpretation (D',δ') *with* D' *countable*, such that $v_{(D',\delta')}(\alpha) = 1$. So, to keep our language fixed, all we need to do is to require it to have a countable set of constants. Admittedly, that won't make *truth under an interpretation* the same under the x-variant and substitutional readings of the quantifier, but it will make *logical consequence, logical equivalence*, and *logical truth*, shortly to be defined, the same – which is what we are really interested in.

Alice: I am afraid that I am rather lost there, but take your word for it. Nonetheless, I must admit that somehow I still have a preference for the x-variant reading. Perhaps I am a mathematician at heart, rather than a philosopher or computer scientist.

EXERCISE 9.3.3 (WITH PARTIAL SOLUTION)

Take the formulae of Exercise 9.3.1 (a), (b), (c) and obtain the same results using the substitutional reading of the quantifiers in the two element domain $\{1, 2\}$.

Solution to (a): Put $\delta(P) = \{1\}$ and $\delta(Q) = \{2\}$, and let a, b be constants with $\delta(a) = 1$ and $\delta(b) = 2$. We claim that $v(\forall x(Px \lor Qx)) = 1$ while $v(\forall x(Px) \lor \forall x(Qx)) = 0$. For the former, it suffices to show that both

$v(Pa \lor Qa) = 1$ and $v(Pb \lor Qb) = 1$. But since $v(Pa) = 1$ we have $v(Pa \lor Qa)$ $= 1$, and likewise since $v(Qb) = 1$ we have $v(Pb \lor Qb) = 1$. For the latter, $v(Pb) = 0$ so that $v(\forall x(Px)) = 0$, and likewise $v(Qa) = 0$ so that $v(\forall x(Qx))$ $= 0$, and so finally $v(\forall x(Px) \lor \forall x(Qx)) = 0$ as desired.

The substitutional account of the quantifiers throws light on the finite transforms of a quantified formula, which we discussed briefly in the preceding section. Let (D, δ) be any interpretation and v the corresponding valuation using the substitutional reading. Then for any universally quantified formula we have $v(\forall x(\alpha)) = 1$ iff $v(\alpha[a/x]) = 1$ for every constant symbol a of the language. If D is finite and $a_1, \ldots, a_n$ are constants naming all its elements, then this holds iff $v(\alpha[a_1/x] \land \ldots \land (\alpha[a_n/x]) = 1$. Likewise, $v(\exists x(\alpha)) = 1$ iff $v(\alpha[a_1/x] \lor \ldots \lor (\alpha[a_n/x]) = 1$.

Thus the truth-value of a formula under an interpretation (D, δ) where D is finite with n elements can also be calculated by a *translation into a quantifier-free formula* (i.e. one with only truth-functional connectives). We translate $\forall x(\alpha)$ and $(\exists x(\alpha)$ into $\alpha[a_1/x] \land \ldots \land \alpha[a_n/x]$ and $\alpha[a_1/x] \lor \ldots \lor \alpha[a_n/x]$ respectively, where $a_1, \ldots, a_n$ are constants chosen to name all elements of D.

EXERCISE 9.3.4 (WITH PARTIAL SOLUTION)

Take again the formulae of Exercise 9.3.1 (a), (b), (c), translate them into quantifier-free formulae for a two-element domain, and obtain the same results once more by assigning truth-values to atomic formulae and evaluating as in propositional logic.

Solution to (a): Let a, b be constants. The translation of $\forall x(Px \lor Qx)$ is $(Pa \lor Qa) \land (Pb \lor Qb)$ while the translation of $\forall x(Px) \lor \forall x(Qx)$ is $(Pa \land Pb) \lor (Qa \land Qb)$. Let v be the propositional assignment that puts e.g. $v(Pa) = 1 = v(Qb)$ and $v(Pb) = 0 = v(Qa)$. Then by truth-tables $v((Pa \lor Qa) \land (Pb \lor Qb)) = 1$ while $v((Pa \land Pb) \lor (Qa \land Qb)) = 0$ as desired.

9.4 Logical Consequence etc

We now have all the apparatus for defining rigorously the notions of logical consequence, equivalence etc for formulae in the language of quantificational logic. In effect, we take the same definitions as for truth-functional logic, and plug in the more elaborate semantics. To reduce clutter, write v for $v_{(D,\delta)}$. The definitions are:

- *A logically implies* β (or: β is a *logical consequence* of A) and we write $A \mathrel{|\!-} \beta$ iff there is no interpretation (D,δ) such that $v(\alpha) = 1$ for all $\alpha \in A$ but $v(\beta) = 0$.

- α is *logically equivalent* to β, and we write $\alpha \mathrel{-\|\!-} \beta$, iff both $\alpha \mathrel{|\!-} \beta$ and $\beta \mathrel{|\!-} \alpha$. Equivalently: $\alpha \mathrel{-\|\!-} \beta$ iff $v(\alpha) = v(\beta)$ for every interpretation (D,δ).

- α is a *logically true* iff $v(\alpha) = 1$ for every interpretation (D,δ).

- α is a *contradiction* iff $v(\alpha) = 0$ for every interpretation (D,δ). More generally, a set A of formulae is *inconsistent* iff for every interpretation (D,δ) there is a formula $\alpha \in A$ with $v(\alpha) = 0$.

- α is *contingent* iff it is neither logically true nor a contradiction.

Evidently, the five bulleted concepts above are interrelated in the same way as their counterparts in propositional logic, with essentially the same verifications. For example, α is logically true iff $\varnothing \mathrel{|\!-} \alpha$.

In Section 9.2 we already noted a number of important logical equivalences, and they can all be verified using the definitions above. For example, the first quantifier interchange principle $\neg\forall x(\alpha) \mathrel{-\|\!-} \exists x(\neg\alpha)$ can be verified in excruciating detail as follows, using say the x-variant reading. Consider any interpretation (D,δ), and let $v = v_{D,\delta}$. Then $v(\neg\forall x(\alpha)) = 1$ iff $v(\forall x(\alpha)) = 0$, iff not every x-variant interpretation (D,δ') gives $v'(\alpha) = 1$, iff some x-variant interpretation (D,δ') gives $v'(\alpha) = 0$, iff some x-variant interpretation (D,δ') gives $v'(\neg\alpha) = 1$, iff $v(\exists x(\neg\alpha)) = 1$ as desired.

EXERCISE 9.4.1

(a) Verify the same logical equivalence using the substitutional reading of the quantifiers.

(b) Verify one of the distribution equivalences of Section 9.2, using either the x-variant or the substitutional reading.

In exercises of Section 9.3 we also checked some negative results. In effect, we saw that $\forall x(Px \vee Qx) \mathrel{|\!\not-} \forall x(Px) \vee \forall x(Qx)$, $\exists x(Px) \wedge \exists x(Qx) \mathrel{|\!\not-} \exists x(Px \wedge Qx)$, and $\forall x\exists y(Rxy) \mathrel{|\!\not-} \exists y\forall x(Rxy)$. In each instance we found an interpretation in a very small domain (two elements) that does the job. For more complex non-implications one often needs larger domains. Indeed there are formulae α, β such that $\alpha \mathrel{|\!\not-} \beta$ but such that $v(\beta) = 1$ for every interpretation in a finite domain with $v(\alpha) = 1$: in other words, such non-implications can be witnessed only in infinite domains (which, however, can always be chosen to be countable). But that is beyond our remit, and we remain with more elementary matters.

The last of the three non-implications above suggests the general question of which alternating quantifiers imply which. This is answered in the following diagram.

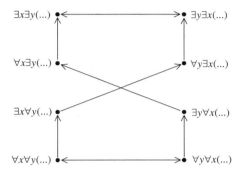

Figure 9.1 *Logical relations between alternating quantifiers.*

The double-headed arrows in the diagram indicate logical equivalence; single-headed ones are for logical implication. We are looking at formulae with two initial quantifiers with attached variables; the left column commutes the quantifiers leaving the attached variables in the fixed order x,y ($2^2 = 4$ cases), while the right column does the same with the attached variables in reverse order y,x (another $2^2 = 4$ cases), thus 8 cases in all. In each case the material quantified remains unchanged. The diagram is complete in the sense that when there is no path from φ to ψ in the figure, then $\varphi \not\models \psi$.

EXERCISE 9.4.2

(a) Verify the non-implication $\exists x \forall y(\alpha) \not\models \exists y \forall x(\alpha)$ by choosing α to be Rxy and considering a suitable interpretation. Make life easy by using the smallest domain that you can get away with. Use either the x-variant or substitutional account of the quantifier in your verification, as you prefer.

(b) Verify the implication $\exists x \forall y(\alpha) \models \forall y \exists x(\alpha)$ using the semantics for the quantifiers. Again, use either the x-variant or substitutional account of the quantifier in your verification. *Warning*: Evidently, it is not enough to consider a single interpretation; you need to show that *every* interpretation that makes the LHS true also makes the RHS true.

(c) Correct or incorrect? (i) $\forall x(Px \to Qx) \mathrel{|-} \forall x(Px) \to \forall x(Qx)$, (ii) the converse.

These relations between alternating quantifiers rest on four fundamental principles for the quantifiers. To formulate them we need a further concept – that of a *clean* substitution.

Recall that a term t may contain variables as well as constants; indeed a variable standing alone is a term. So when we substitute t for free occurrences of x in α, it may happen that t contains a variable y that is 'captured' by some quantifier of α. For example, when we substitute y for the sole free occurrence of x in $\exists y(Rxy)$ we get $\alpha[y/x] = \exists y(Ryy)$: the y that is introduced is in the scope of the existential quantifier. Such substitutions are undesirable from the point of view of the principles that we are about to articulate. We say that a substitution $\alpha[t/x]$ is *clean* iff no free occurrence of x in α falls within the scope of a quantifier binding some variable occurring in t.

EXERCISE 9.4.3 (WITH SOLUTION)

Let $\alpha = \exists y(Py \lor \forall z(Rzx \land \forall x(Sax)))$. Which of the following substitutions are clean? (i) $\alpha[y/x]$, (ii) $\alpha[a/x]$, (iii) $\alpha[x/x]$, (iv) $\alpha[z/x]$, (v) $\alpha[z/y]$, (vi) $\alpha[y/w]$.

Solution: For the first four, note that there is just one free occurrence of x in α. Then: (i) Not clean, because this occurrence of x is in the scope of $\exists y$. (ii), (iii) Both clean. (iv) No, because the free occurrence of x is in the scope of $\forall z$. For (v), note that there are no free occurrences of y. Hence, vacuously, the substitution $\alpha[z/y]$ is clean. (vi) Again vacuously clean: there are no occurrences at all of w.

Comment: This exercise illustrates the following facts, immediate from the definition: (a) The substitution of a constant for a variable is always clean, (b) The identity substitution is always clean, (c) A substitution is clean whenever the variable being replaced has no free occurrences in the formula – a *fortiori* when it has no occurrences at all.

We may now formulate the promised principles about $\forall$. The first is a logical consequence, the second is a rule about logical consequences.

- $\forall-$: $\forall x(\alpha) \mathrel{|-} \alpha[t/x]$, provided the substitution $\alpha[t/x]$ is clean.

- $\forall+$: Whenever $\alpha_1, \ldots, \alpha_n \mathrel{|-} \alpha$ then $\alpha_1, \ldots, \alpha_n \mathrel{|-} \forall x(\alpha)$, provided the variable x has no free occurrences in any of $\alpha_1, \ldots, \alpha_n$.

As you would expect, there are two dual principles about $\exists$:

- $\exists+$: $\alpha[t/x]$ $\vdash$ $\exists x(\alpha)$, provided the substitution $\alpha[t/x]$ is clean.

- $\exists-$: Whenever $\alpha_1,\ldots,\alpha_{n-1}, \alpha \vdash \alpha_n$ then $\alpha_1,\ldots,\alpha_{n-1}, \exists x(\alpha) \vdash \alpha_n$, provided the variable x has no free occurrences in any of $\alpha_1,\ldots,\alpha_n$.

EXERCISE 9.4.4 (WITH PARTIAL SOLUTION)

(a) Which of the following are instances of $\forall-$? (i) $\forall x(Px{\rightarrow}Qx) \vdash Px{\rightarrow}Qx$, (ii) $\forall x(Px{\rightarrow}Qx) \vdash Pb{\rightarrow}Qb$, (iii) $\forall x(Px{\wedge}Qx) \vdash Px{\wedge}Qy$, (iv) $\forall x\exists y(Rxy) \vdash \exists y(Rxy)$, (v) $\forall x\exists y(Rxy) \vdash \exists y(Ryy)$, (vi) $\forall x\exists x(Rxy) \vdash \exists x(Ray)$, (vii) $\forall x\exists y(Rxy) \vdash \exists y(Ryx)$, (viii) $\forall x\exists y(Rxy) \vdash \exists y(Rzy)$, (ix) $\forall x\exists y(Rxy) \vdash \exists y(Ray)$, (x) $\forall x\exists y(Rxy) \vdash Rxy$.

(b) Which of the following are instances of $\exists+$? (i) $Pa \vdash \exists x(Px)$, (ii) $Rxy \vdash \exists x(Rxx)$, (iii) $Rxy \vdash \exists z(Rzy)$, (iv) $Rxy \vdash \exists z(Rzx)$, (v) $\exists y(Rxy) \vdash \exists z\exists y(Rzy)$.

(c) Which of the following are justified by the indirect rule $\forall+$, given that $\forall x\exists y(Rxyz) \vdash \exists y(Rzyz)$? (i) $\forall x\exists y(Rxyz) \vdash \forall z\exists y(Rzyz)$, (ii) $\forall x\exists y(Rxyz) \vdash \forall w\exists y(Rzyz)$, (iii) $\forall x\exists y(Rxyz) \vdash \forall w\exists y(Rwyz)$, (iv) $\forall x\exists y(Rxyz) \vdash \exists x\exists y(Rzyz)$, (v) $\forall x\exists y(Rxyz) \vdash \forall y\exists y(Rzyz)$.

(d) Which of the following are justified by the indirect rule $\exists-$, given that $Px,\forall y(Ryz) \vdash Px{\wedge}\exists z(Rxz)$? (i) $Px, \exists z\forall y(Ryz) \vdash Px{\wedge}\exists z(Rxz)$, (ii) $Px, \exists x\forall y(Ryz) \vdash Px{\wedge}\exists z(Rxz)$, (iii) $Px, \exists w\forall y(Ryw) \vdash Px{\wedge}\exists z(Rxz)$, (iv) $Px, \exists w\forall y(Ryz) \vdash Px{\wedge}\exists z(Rxz)$, (v) $\exists x(Px),\forall y(Ryz) \vdash Px{\wedge}\exists z(Rxz)$.

Solution to (a): (i), (ii) Yes; (iii) no; (iv) yes; (v), (vi) and (vii) no; (viii) and (ix) yes; (x) no.

Note carefully that the two direct implications $\forall-$ and $\exists+$ make substitutions and require cleanliness. In contrast, the indirect ones $\forall+$ and $\exists-$ do not make substitutions, and have the different proviso that the quantified variable has no free occurrences in certain formulae. Commit to memory to avoid conflation!

Alice Box: Why the minus in $\exists-$?

Alice: I'm OK with all that, but I'm bothered by the notation. I can see why you label the first three rules as you do, but not the fourth. Why the minus sign in $\exists-$? Surely we are *adding* the existential quantifier $\exists x$ to α!

(*Continued*)

Alice Box: (Continued)

Hatter: The name ∃– comes from the use of this principle in natural deduction. We will come to that shortly, but I think I can explain already the *raison d'être* for the minus sign. If we want to establish the principal consequence it suffices to show the subordinate one. In other words, given the premises $\alpha_1, \ldots, \alpha_{n-1}, \exists x(\alpha)$ we can *strip off* the existential quantifier, use α as an additional premise, and head for the same goal α_n. We are not *inferring* α from $\exists x(\alpha)$, but *procedurally* we are moving from $\exists x(\alpha)$ to α in building our indirect inference.

So far we have ignored the identity relation sign. Recall from the preceding section that this symbol is always given a highly constrained interpretation. At the most lax, it is interpreted as a congruence relation; in the standard semantics (used here) it is always taken as the identity relation over the domain of discourse. Whichever of these interpretations we follow, standard or non-standard, two special logical implications hold.

The first one reflects the fact that identity is a reflexive relation. It tells us that the formula $\forall x(x \equiv x)$ is true under all interpretations. It will be convenient for us to express it as a logical consequence with the empty set of premises:

$$\varnothing \mid\!- \forall x(x \equiv x)$$

The second principle for identity reflects the fact that whenever x is identical with y then whatever is true of x is true of y. Sometimes dubbed the principle of the 'indiscernibility of identicals', its formulation in the language of first-order logic is a little trickier than one might anticipate.

We need the concept of the *replacement* of one term by another. Let t, t' be terms. They may be constants, variables, or more complex terms made up from constants and/or variables by iterated application of function letters. We have already defined what it means for an occurrence of a *variable* to be free in a formula; now generalize to say that the *term t is free* in α iff no occurrence of any variable in t falls within the scope of a quantifier of α with the same variable attached to it. Write $\alpha[t'/\!/t]$ (with two strokes rather than one) to indicate the result of taking any one free occurrence of the term t in α and replacing it by t'. The replacement $\alpha[t/\!/t']$ is said to be *clean* if the occurrence of t' introduced is free in $\alpha[t/\!/t']$. Then we have the following principle of the *indiscernibility of identicals*:

$\alpha, t \equiv t'$ |- $\alpha[t'/\!/t]$, provided the replacement is clean.

For example, taking α to be $\exists y(Rxy)$, t to be x, and t' to be z we have:

$$\exists y(Rxy), x \equiv z\ |\text{-}\ \exists y(Rzy),$$

since the introduced variable z is free in $\exists y(Rzy)$. On the other hand, if we take t' to be y then it is *not* free in $\alpha[t'/\!/t]$, ie. in $\exists y(Ryy)$, and so the principle *does not authorize* the consequence $\exists y(Rxy)$, $x \equiv y$ |- $\exists y(Ryy)$. Intuitively speaking, the y free in the formula $x \equiv y$ has nothing to do with the y bound in $\exists y(Ryy)$. The latter is logically equivalent to $\exists z(Rzz)$, and clearly $\exists y(Rxy)$, $x \equiv y$ |$\not$- $\exists z(Rzz)$.

EXERCISE 9.4.5 (WITH SOLUTION)

Put α to be $\exists y(Rf(x,a),y)$ and t to be $f(x,a)$. Let t_1 be $g(x)$, and t_2 be $f(y,z)$. Write out $\alpha[t_1/\!/t]$ and $\alpha[t_2/\!/t]$, and determine in each case whether the principle authorizes the consequence α, $t \equiv t_i$ |- $\alpha[t_i/\!/t]$.

Solution: With t_1 chosen as $g(x)$, $\alpha[t_1/\!/t]$ becomes $\exists y(g(x),y)$, in which the introduced $g(x)$ is free. The principle thus gives us $\exists y(Rf(x,a),y)$, $f(x,a) \equiv g(x)$ |- $\exists y(Rg(x),y)$. On the other hand, with t_2 chosen as $f(y,z)$, $\alpha[t_1/\!/t]$ becomes $\exists y(Rf(y,z),y)$, in which the introduced $f(y,z)$ is bound. The principle *does not authorize* the implication $\exists y(Rf(x,a),y)$, $f(x,a) \equiv f(y,z)$ |- $\exists y(Rf(y,z),y)$.

The proof of the principle of the indiscernibility of identicals is by a rather tedious induction on the complexity of the formula α, which we omit.

9.5 Natural Deduction with Quantifiers

Can we extend, to the language of the quantifiers, the shortcut systems that were so successful in propositional logic? The short answer is *yes*, although there are nuances.

In the case of *semantic decomposition trees*, we can introduce rules for decomposing the quantifiers and handling identity, but the system no longer gives us a decision procedure. There is no guarantee that the tree terminates in a finite time: it may have infinitely long branches. Some of the infinitely long branches can be avoided by allowing the decomposition of an existential statement to give rise to many children; but some infinite branches (corresponding to interpretations in infinite domains) may still remain. For them, control procedures are also needed to ensure that every decomposition rule eventually gets applied, i.e. that it is not postponed indefinitely while another rule is applied infinitely many times. Finally, if we allow non-closed formulae, the identity symbol and function letters in the language, then the exact formulation of the decomposition rules becomes rather intricate. Thus, although the study of first-order decomposition trees is fascinating, it is best left to a course devoted to first-order logic.

On the other hand, natural deduction did not give a decision procedure even in the limited context of propositional logic, and the transition to the quantificational context is rather smoother. We describe it briefly in this section. In effect, we take the six rules for quantifiers and identity that are presented above, and add them to the stock available from propositional logic. The following four examples illustrate how this is done.

Example 1. Construct a derivation for the logical consequence $\forall x(Px \rightarrow Qx)$ |- $\forall x(Px) \rightarrow \forall x(Qx)$, using the rules $\forall-$ and $\forall+$.

Table 9.6 A natural deduction using rules for the universal quantifier

N°	Formula	From	Rule	Depends on	Current Goal
1	$\forall x(Px \rightarrow Qx)$		premise	1	$\forall x(Px) \rightarrow \forall x(Qx)$
2	$\forall x(Px)$		supposition	2	$\forall x(Qx)$
3					Qx
4	Px	2	$\forall-$, proviso OK	2	ditto
5	$Px \rightarrow Qx$	1	$\forall-$, proviso OK	1	ditto
6	Qx	4,5	modus ponens	1,2	$\forall x(Qx)$
7	$\forall x(Qx)$	1,2 |- 6	$\forall+$, proviso OK	1,2	$\forall x(Px) \rightarrow \forall x(Qx)$
8	$\forall x(Px) \rightarrow \forall x(Qx)$	1,2 |- 7	conditional proof (discharge 2)	1	□

Commentary. As the conclusion is a conditional, we begin with the supposition $\forall x(Px)$ for a conditional proof, the goal becoming $\forall x(Qx)$. Noting that the variable x has no free occurrences in either the initial premise or the supposition, we can reset our goal to be just Qx. We are not applying $\forall+$ yet: we are *initiating a strategy*, foreseeing that we should be able to apply it later. Nor are we making any supposition – the principal and the subordinate inferences for the rule $\forall+$ have exactly the same premises, in contrast with all of the other indirect rules considered. For this reason, the left columns of row 3 are left blank. Straightforward applications of $\forall-$ and modus ponens lead us to Qx as desired, at which point our goal reverts to $\forall x(Qx)$. This we get by the awaited application of $\forall+$ (with $n = 2$), noting that the proviso is satisfied (the variable x has no free occurrences in the formulae 1,2 on which Qx depends). The final application of conditional proof completes the derivation.

A lot of analysis for a simple derivation! But that is logic. The pattern revealed here is quite typical. Roughly speaking, we strip off quantifiers using $\forall-$, make truth-functional inferences, and finally use ∀+ to get the quantifiers back in a last-off/first-on fashion. Hence the symmetry of the 'current goal' column.

Example 2. The second example shows how the same striptease may be carried out with the existential quantifier. We construct a derivation for the logical consequence $\exists x(Px \wedge Qx)$ |- $\exists x(Px) \wedge \exists x(Qx)$, using the rules $\exists$– and $\exists$+.

Table 9.7 A natural deduction using rules for the existential quantifier

N°	Formula	From	Rule	Depends on	Current Goal
1	$\exists x(Px \wedge Qx)$		premise	1	$\exists x(Px) \wedge \exists x(Qx)$
2	$Px \wedge Qx$		supposition	2	ditto
3	Px	2	simplification	2	ditto
4	Qx	2	simplification	2	ditto
5	$\exists x(Px)$	3	$\exists$+, proviso OK	2	ditto
6	$\exists x(Qx)$	4	$\exists$+, proviso OK	2	ditto
7	$\exists x(Px) \wedge \exists x(Qx)$	5, 6	conjunction	2	ditto
8	$\exists x(Px) \wedge \exists x(Qx)$	2 |- 7	$\exists$–, proviso OK (discharge 2)	1	□

Commentary. Observing that the variable x has no free occurrences in either the initial premise or the conclusion, we make the supposition $Px \wedge Qx$, planning to end up with an application of $\exists$–. Note that the goal remains unchanged as $\exists x(Px) \wedge \exists x(Qx)$, following the pattern of the rule for $\exists$–. We are not applying $\exists$– yet: once again, we are initiating a strategy to apply it later. Straightforward applications of $\exists$+ and truth-functional steps lead us to $\exists x(Px) \wedge \exists x(Qx)$ as desired, but still depending on the supposition 2. Applying finally $\exists$– as planned (with $n = 1$) makes the same conclusion depend on 1, as desired.

Example 3. Our third example combines the rules for the two quantifiers. We construct a derivation for the logical consequence $\exists x \forall y(Rxy)$ |- $\forall y \exists x(Rxy)$, one of the principles concerning alternating quantifiers that were displayed in Figure 9.1. As there are no truth-functional connectives in this, the derivation will be a purely quantificational one.

Table 9.8 A natural deduction using rules for both quantifiers

N°	Formula	From	Rule	Depends on	Current Goal
1	$\exists x \forall y(Rxy)$		premise	1	$\forall y \exists x(Rxy)$
2	$\forall y(Rxy)$		supposition	2	ditto
3					$\exists x(Rxy)$
4	Rxy	2	$\forall$–, proviso OK	2	ditto
5	$\exists x(Rxy)$	4	$\exists$+, proviso OK	2	$\forall y \exists x(Rxy)$
6	$\forall y \exists x(Rxy)$	2 |- 5	$\forall$+, proviso OK	2	ditto
7	$\forall y \exists x(Rxy)$	2 |- 6	$\exists$–, proviso OK discharge 2	1	□

Commentary. Noting that the variable x has no free occurrences in either the given premise or the conclusion, we make the supposition $\forall y(Rxy)$, planning to end up with an application of $\exists-$. Note that the goal remains unchanged as $\forall y\exists x(Rxy)$, in accord with the pattern of the rule for $\exists-$. Observing that the variable y has no free occurrences in 2, we reset the goal to $\exists x(Rxy)$ planning to make a subsequent application of $\forall+$. Straightforward applications of $\forall-$ and $\exists+$ give us that goal, so we apply $\forall+$ as foreseen, getting the initial goal $\forall y\exists x(Rxy)$, but still depending on the supposition 2. Finally applying $\exists-$ as planned (with $n = 1$) makes the same conclusion depend on 1, as desired.

Clearly, in this example, some freedom is possible in the order in which things are done. In particular, we could have written row 3 (reset goal) before row 2 (supposition). But care should be taken: ill-considered reordering can violate the provisos on the quantifier rules.

By now it is evident that in order to construct derivations using indirect rules, whether propositional or quantificational, one needs to *plan ahead*. Near the beginning we make moves (such as stripping off quantifiers or making suppositions) *in order to* make other moves (discharging suppositions and restoring quantifiers) near the end.

Example 4. Our last example uses the rules for identity. We construct a derivation for the logical consequence $\varnothing \mid- \forall x\forall y((x \equiv y){\rightarrow}(y \equiv x))$, which expresses the symmetry of identity.

Table 9.9 A natural deduction using rules for identity

N°	Formula	From	Rule	Current Goal
1				$\forall x\forall y((x \equiv y){\rightarrow}(yx))$
2				$\forall y((x \equiv y){\rightarrow}(y \equiv x))$
3				$(x \equiv y){\rightarrow}(y \equiv x)$
4	$x \equiv y$		supposition	$y \equiv x$
5	$\forall x(x \equiv x)$		first identity rule	ditto
6	$x \equiv x$	5	$\forall-$, proviso OK	ditto
7	$y \equiv x$	6, 4	second identity rule	$(x \equiv y){\rightarrow}(y \equiv x)$
8	$(x \equiv y){\rightarrow}(y \equiv x)$	4 $\mid- 6$	conditional proof	$\forall y((x \equiv y){\rightarrow}(y \equiv x))$
9	$\forall y((x \equiv y){\rightarrow}(y \equiv x))$	$\varnothing \mid- 8$	$\forall+$, proviso OK	$\forall x\forall y((x \equiv y){\rightarrow}(y \equiv x))$
10	$\forall x\forall y((x \equiv y){\rightarrow} (y \equiv x))$	$\varnothing \mid- 9$	$\forall+$, proviso OK	$\square$

Commentary. Because of limited page width we have omitted the "depends on" column. Note that the formulae in rows 4 and 7 depend on supposition 4; the other formulae depend on nothing. We focus on the aspects that concern identity. We are trying to get $\forall x \forall y((x \equiv y) \to (y \equiv x))$ from no premises at all; hence the empty cells in row 1. We plan to end the derivation with a double $\forall+$, hence rows 2 and 3. That suggests a conditional proof, whose supposition is in row 4. We can place $\forall x(x \equiv x)$ in row 5 because we know that it follows from the empty set of premises; and we then strip off its quantifier in row 6. This brings us to the *key step of the whole derivation* in row 7, which is an application of the second identity rule. How? We are claiming that 6, 4 |- 7, i.e. that $x \equiv x$, $x \equiv y$ |- $y \equiv x$. This is of the form α, $t \equiv t'$ |- $\alpha[t'/\!/t]$, where α is $x \equiv x$, t is x, t' is y, and we replace the *first* occurrence of x in α by y, giving us $\alpha[t'/\!/t]$ which is $y \equiv x$. This replacement preserves freedom, so the proviso of the rule is satisfied. The rest of the derivation discharges the supposition and restores quantifiers.

A rather intricate derivation for such a simple fact! To get things in perspective, see row 7 as the crux of the procedure, with the preceding and following rows as stripping off then getting dressed again.

EXERCISE 9.5.1

Carry out natural deductions for each of the following logical consequences.

(a) $\forall x(Px)$ |- $\exists y(Py)$

(b) $\exists x(Px)$ |- $\exists y(Py)$

(c) $\forall x(Px \wedge Qx)$ |- $\forall x(Px) \wedge \forall y(Qy)$

(d) $\exists x \forall y(Rxy)$ |- $\exists x(Rxx)$

The rules of natural deduction that we have introduced in this chapter and the preceding one are *sound*, in the sense that whenever there is some derivation (of finite length) of a formula α from a set A of formulae, using the rules of enchainment and indirect inference that we have presented, then A |- α. The proof is essentially by induction on the length of the derivation, using the fact that all of the enchainment rules are sound, and that all of the indirect rules preserve logical consequence.

Moreover, the converse is true: it can be shown that the system of natural deduction is *complete* for quantificational logic. That is to say, whenever A |- α there is some derivation (of finite length) of α from A using the rules of enchainment and indirect inference that we have described in this chapter.

These two theorems, soundness and completeness, stand at the door of an advanced study of the subject, beyond what is appropriate for an introductory course, like this one, on sets, logic and finite mathematics.

FURTHER EXERCISES

9.1. *The language of quantifiers*

(a) Return to the questions raised in the second part of Table 9.1 and consider them again in the light of the chapter.

(b) Express the following in the language of first-order logic, in each case specifying explicitly an appropriate domain of discourse and suitable constants, function letters, and/or predicates:

   (i) Zero is less than or equal to every natural number

   (ii) If one real number is less than another, there is a third between the two

   (iii) Every computer program has a bug

   (iv) Any student who can solve this problem can solve all problems

   (v) Squaring on the natural numbers is injective but is not onto

   (vi) Nobody loves everybody, but nobody loves nobody.

(c) Identify the free and bound occurrences of variables and terms in the formula $\alpha = \forall x \exists y ((y \equiv z) \rightarrow \forall w \exists y (Rf(u,w),y))$. Identify also the vacuous quantifiers, if any. Which of the substitutions $\alpha[x/z]$, $\alpha[y/z]$, $\alpha[w/z]$, and their three converse substitutions, are clean?

9.2. *Interpretations*

(a) Consider the relation between interpretations of being x-variants (for a fixed variable x). Is it an equivalence relation? And if the variable x is not held fixed? Justify.

(b) Find a formula using the identity predicate, which is true under every interpretation whose domain has less than three elements, but is false under every interpretation whose domain has three or more elements.

(c) Sketch an explanation why there is no hope of finding a formula that does not contain the identity predicate, with the same properties.

(d) In the definition of an interpretation, we required that the domain of discourse is non-empty. Would the definition make sense, as it stands, if we allowed the empty domain? How might you reformulate it if you wanted to admit the empty domain? What would the result be for quantified statements?

9.3. *Logical consequence etc*

(a) Justify the principle of vacuous quantification semantically, first using the x-variant reading of the quantifiers, and then using the substitutional reading. Which is easier?

(b) Which of the following claims are correct? (i) $\forall x(Px \to Qx)$ $\vdash \exists x(Px) \to \exists x(Qx)$, (ii) $\exists x(Px \to Qx)$ $\vdash$ $\exists x(Px) \to \exists x(Qx)$, (iii) $\forall x(Px \to Qx)$ $\vdash \exists x(Px) \to \forall x(Qx)$, (iv) $\forall x(Px \to Qx)$ $\vdash \forall x(Px) \to \exists x(Qx)$, and (v) – (viii) their converses. In the negative cases, give an interpretation in a small finite domain to serve as witness. In the positive cases, sketch a semantic verification using either the x-variant or the substitutional reading of the quantifiers.

(c) Find a simple formula α such that $\forall x(\alpha)$ is a contradiction but α is not. Use this to get a simple formula β such that $\exists x(\beta)$ is logically true but β is not. *Hint*: You can do it without the identity predicate.

(d) Recall from the text that a set A of formulae is said to be *consistent* iff there is some interpretation that makes all of its elements true. Verify the following (i) The empty set is consistent, (ii) A singleton set is consistent iff its unique element is not a contradiction, (iii) An arbitrary set A of formulae is consistent iff $A \not\vdash p \land \neg p$. (iv) Every subset of a consistent set of formulae is consistent, (v) A finite set of formulae is consistent iff the conjunction of all its elements is consistent.

9.4. *Natural deduction*

Construct natural deductions for the following logical consequences.

(a) $\exists x(Px) \vdash \forall x \exists x(Px)$.

(b) $\varnothing \vdash \forall x \forall y \forall z((x \equiv y \land y \equiv z) \to (x \equiv z))$. *Hint*: Strip off the quantifiers, then use conditional proof, within which you will need to use the second principle for identity.

(c) $\forall x \forall y(Rxy) \vdash \forall x \forall y(Ryx)$. *Hint*: Your first attempt may run into a substitution snag. Try introducing a fresh variable z and then getting rid of it.

(d) $\varnothing \vdash \neg \exists y \forall x(Rxy \leftrightarrow \neg Rxx)$. *Remark*: Readers familiar with the paradoxes of naïve set theory will see this as reflecting Russell's paradox in abstract form, with an arbitrary relation symbol R replacing $\in$.

Those unfamiliar with such matters can still do the exercise. Use proof by contradiction and strip off.

9.5. *General reflection*

(a) Given that the method of natural deduction is complete for quantificational logic, sketch an argument to show that whenever an infinite set of formulae is inconsistent, it must have an inconsistent finite subset. *Hint*: First relate inconsistency to logical consequence as in Exercise 9.3(d)(iii), then make use of the fact that every derivation is, by definition, of finite length.

(b) Give an intuitive explanation of why the completeness of our system of natural deduction for quantificational logic does not imply its decidability.

Selected Reading

Same books as for propositional logic, with different chapters. We recall that the first book is intended for students of computer science without much mathematics, the next two for the same but with more mathematical sophistication, while the following two are aimed at students of philosophy and the general reader. We add a last reference which, despite its title, may well be used to go into a more advanced study of logic.

James Hein *Discrete Structures, Logic and Computability*. Jones and Bartlett, 2002 (second edition), Chapter 7.

Michael Huth and Mark Ryan *Logic in Computer Science*. Cambridge University Press, 2000, Chapter 4.

Mordechai Ben-Ami *Mathematical Logic for Computer Science*. Springer, 2001 (second edition), Chapters 5–7.

Colin Howson *Logic with Trees*. Routledge, 1997, Chapters 5–11.

Wilfrid Hodges *Logic*. Penguin, 1977, Sections 34–41.

Shawn Hedman *A First Course in Logic*. Oxford, 2004.

Index

Printed in the United States